Zoophysiology   Volume 18

Coordinating Editor: D. S. Farner

Editors:
W. Burggren   S. Ishii   K. Johansen
H. Langer   G. Neuweiler   D. J. Randall

Zoophysiology

Volumes already published in the series:

Volume 1: *P. J. Bentley*
Endocrines and Osmoregulation

Volume 2: *L. Irving*
Arctic Life of Birds and Mammals

Volume 3: *A. E. Needham*
The Significance of Zoochromes

Volume 4/5: *A. C. Neville*
Biology of the Arthropod Cuticle

Volume 6: *K. Schmidt-Koenig*
Migration and Homing in Animals

Volume 7: *E. Curio*
The Etology of Predation

Volume 8: *W. Leuthold*
African Ungulates

Volume 9: *E. B. Edney*
Water Balance in Land Arthropods

Volume 10: *H.-U. Thiele*
Carabid Beetles in Their Environments

Volume 11: *M. H. A. Keenleyside*
Diversity and Adaptation in Fish Behaviour

Volume 12: *E. Skadhauge*
Osmoregulation in Birds

Volume 13: *S. Nilsson*
Autonomic Nerve Function in the Vertebrates

Volume 14: *A. D. Hasler, A. T. Scholz*
Olfactory Imprinting and Homing in Salmon

Volume 15: *T. Mann*
Spermatophores

Volume 16: *P. Bouverot*
Adaption to Altitude-Hypoxia in Vertebrates

Volume 17: *R. J. F. Smith*
The Control of Fish Migration

Volume 18: *E. Gwinner*
Circannual Rhythms

Eberhard Gwinner

Circannual Rhythms

Endogenous Annual Clocks
in the Organization of Seasonal Processes

With 73 Figures

Springer-Verlag
Berlin Heidelberg New York
London Paris Tokyo

Professor Dr. EBERHARD GWINNER
Max-Planck-Institut
für Verhaltensphysiologie, Vogelwarte
8138 Andechs, FRG

ISBN 3-540-16891-5 Springer-Verlag Berlin Heidelberg New York
ISBN 0-387-16891-5 Springer-Verlag New York Berlin Heidelberg

Library of Congress Cataloging in Publication Data. Gwinner, E. (Eberhard), 1938–. Circannual rhythms. (Zoophysiology; v. 18) Includes index. 1. Biological rhythms. I. Title. II. Title: Annual clocks in the organization of seasonal processes. III. Series. QH527.G94 1986 574.5'43 86-17660

This work is subject to copyright. All rights are reserved, whether the whole or part of the material is concerned, specifically those of translation, reprinting, re-use of illustrations, broadcasting, reproduction by photocopying machine or similar means, and storage in data banks. Under § 54 of the German Copyright Law, where copies are made for other than private use, a fee is payable to "Verwertungsgesellschaft Wort", Munich.

© Springer-Verlag Berlin Heidelberg 1986
Printed in Germany

The use of registered names, trademarks, etc. in this publication does not imply, even in the absence of a specific statement, that such names are exempt from the relevant protective laws and regulations and therefore free for general use.

Typesetting, printing and bookbinding: Brühlsche Universitätsdruckerei, Giessen
2131/3130-543210

Preface

In addition to the more or less static properties of the environment, plants and animals must cope with its temporal variations. Among the most conspicuous temporal changes to which organisms are exposed are periodic phenomena generated by the rotation of the earth about its axis, its revolution around the sun, and the more complex movements of the moon in relation to both sun and earth. The first two of these astronomical cycles are basic to the familiar daily and annual rhythms, respectively, in the environment. The third generates somewhat more complex cycles, such as those in moonlight and variations in tides. These environmental cycles have provided challenges and opportunities for organisms to adjust their physiology and behavior to them. Indeed, the predictability inherent to these periodic processes has enabled organisms to evolve innate endogenous rhythmic programs that match the environmental cycles and allow, in a variety of different ways, adjustment of biological activities to the cycles of environmental changes.

The endogenous nature of rhythmicity was first clearly recognized in the 1930's in daily periodicities, the most widely distributed and best investigated class of biological rhythms of this type. In the 1950's, demonstrations of endogenous tidal and lunar rhythms, which occur in some littoral and marine organisms, ensued. Another decade passed before endogenous annual periodicities were first demonstrated unambiguously. The reason for this delay is due mainly to the difficulties in experimental demonstration of the spontaneous nature of rhythms with a cycle duration of as long as about a year. As Menaker (1974) stated: "Perhaps the major difficulty in the study of circannual rhythms is a consequence of the ratio of the period length of a single circannual cycle to the length of the productive life of the biologist." Despite this frustrating fact, the phenomenon of circannual rhythmicity attracted a substantial corps of investigators during the ensuing decade. The long-term nature of the processes involved creates not only frustration but also fascination. The major purpose of this book is to promote the latter!

In spite of recent progress, research on endogenous circannual rhythms is still in a rather undeveloped state. In such a situation an inventory of the facts available should be helpful.

Therefore in the first four chapters of this book I have accumulated all the relevant data that I could find. I have attempted to arrange these data in meaningful order and to extract generalizations from them. Here I allowed myself to be led in part by the developments that had taken place in the field of circadian rhythms; hence the recurring comparisons between properties of circannual and circadian rhythms. I have explored the historical roots of research on circannual rhythms by study of early publications on the control of annual rhythmicity. This proved to be a fascinating enterprise, as I found that the literature contains a significant number of early suggestions and descriptions of ingenious experiments on the subject. Since a number of these older papers were written in German, my native language, I found it particularly appropriate to summarize and critically evaluate these results in English.

The most difficult part of the book to prepare was Chapter 5, on mechanisms. Although it has been rewritten and reorganized several times, I remain dissatisfied with it. The problems stem partly from our almost complete ignorance of the physiological processes involved in generating circannual rhythmicity. Moreover, the few available analyses of mechanisms have been attempted at rather different levels of organization and it is difficult to "translate" results obtained at one level to another. Finally, organisms from very different taxa have been studied and it is likely that the circannual clocks in different groups are controlled by rather different mechanisms. I nevertheless hope that the chapter will be heuristic, at least in stimulating criticism and, thereby, progress.

The treatise on adaptive functions in Chapter 6 is my favorite because it presents the aspects of endogenous circannual rhythms that has absorbed most of my research time over the past two decades. Although I have attempted to avoid an excessive influence of my enthusiasm about this subject, I am aware that this chapter is still biased toward my own investigations and those of my colleagues.

Many people have helped in the preparation of this book. Professor Donald S. Farner initially suggested writing it. Apart from giving valuable advice about improving its content, he edited the manuscript thoroughly and revised the English. Dr. John Dittami corrected the English of an early version of the manuscript. He was always prepared for discussions about critical issues and more than once suggested important changes and improvements. Professor Jürgen Aschoff read the manuscript and helped with critical and stimulating comments. More importantly, he was the one who initially evoked my interest in biological rhythms and who kept this interest alive during the 15 years I worked at his institute and thereafter. His classical paper *Jahresperiodik der Fortpflanzung bei Warmblütern* provided the initial stimulus for my work on annual rhythmicities. The comments and critical evaluation of Professor Nicholas Mrosovsky, who read an early draft of the entire book, were of invaluable help. The discussions with him about crucial issues have helped me very extensively in clarifying my own thoughts. Professor Erwin Bünning kindly read the sections on circannual rhythms in plants and made significant suggestions. I would also like to thank Mrs. Ingrid Schwabl for preparing most of the drawings and for her assistance in many of the experiments portrayed in this book. Last but not least, I am grateful to my wife Helga for her support and encouragement, as well as for preparing the design for the front cover illustration.

Andechs, Summer 1986   E. GWINNER

Contents

Chapter 1. *Introduction* 1

1.1 The Phenomenon of Annual Rhythmicity 1
1.2 Ultimate and Proximate Factors in the Control of
    Annual Rhythms 1
    1.2.1 Ultimate Factors 1
    1.2.2 Proximate Factors 2
1.3 Circannual Rhythms 4
1.4 Hierarchical Organization of Proximate Factors . . 4
1.5 Early Suggestions of Circannual Rhythms 5
    1.5.1 Mammalian Hibernation and Avian Migration 5
    1.5.2 Yearly Breeding Cycles in Tropical and
          Temperate-Zone Animals 6
    1.5.3 Ten-Month Breeding Cycles in Tropical Birds 7
    1.5.4 Plants 7
    1.5.5 Development of the Oscillator Analogy . . 8
1.6 Some Definitions and Delimitations 8

Chapter 2. *Evidence for Circannual Rhythms* 11

2.1 Typical Cases 11
    2.1.1 Mammals 11
    2.1.2 Birds 24
    2.1.3 Lower Vertebrates 28
    2.1.4 Invertebrates 29
    2.1.5 Plants 31
2.2 Atypical Cases 35

Chapter 3. *Properties of Free-Running Circannual
           Rhythms* 39

3.1 Degree of Persistence and Range of Permissive
    Conditions 39
3.2 Range of Circannual Period Length and Transients 41
3.3 Dependence of Period on External Conditions . . 42
    3.3.1 Temperature 42
    3.3.2 Photoperiod 44
    3.3.3 Social Factors 47

3.4 Innateness of Circannual Rhythms 47
3.5 Comparison with Circadian Rhythms 47

Chapter 4. *Synchronization of Circannual Rhythms* . . . 49

4.1 Zeitgebers 49
    4.1.1 Photoperiod 50
    4.1.2 Ambient Temperature 58
    4.1.3 Social Stimuli 60
4.2 Ranges of Entrainment 61
4.3 Behavoir Within the Range of Entrainment 63
4.4 Comparison with Circadian Rhythms and Some
    Conclusions 67

Chapter 5. *Mechanisms of Circannual Organization* . . 69

5.1 Interactions with the Circadian System 69
    5.1.1 Frequency Demultiplication of Circadian
        Rhythms 70
        5.1.1.1 Relationship Between Circadian and
              Circannual Period Length 70
        5.1.1.2 Effects of Disrupting the Circadian
              System on Circannual Rhythmicity . . 74
        5.1.1.3 General Properties of Circannual
              Rhythms in Conflict with the Model . 75
    5.1.2 Circannual Rhythm of a Circadian Rhythm
        in Photosensitivity 76
    5.1.3 Circannual Variations in the Internal Circadian
        System 78
        5.1.3.1 Relationship Between States of
              Circadian and Circannual System . . 79
        5.1.3.2 Internal Coincidence Between Circadian
              Neurotransmitter Rhythms as the Basis
              of Circannual Changes? 82
5.2 Interrelationship Among Different Circannual
    Functions. Or: One or Several Circannual Clocks? . 82
    5.2.1 Differential Degrees of Persistence of Various
        Circannual Functions 84
    5.2.2 Internal Dissociation of Various Circannual
        Functions 85
    5.2.3 Selective Manipulation of Rhythmic Functions 88
    5.2.4 Conclusions 90
5.3 Components of Specific Circannual Functions . . . 90
    5.3.1 Attempts at Changing Circannual Period by
        Altering the Duration of Potential Components
        of the Cycle 90

5.3.2 Attempts at Identifying External and Internal Conditions Under Which Rhythmicity Stops or Continues 91
5.3.3 Identification of Hormones and Central Nervous Structures Involved in the Control of Circannual Rhythms 93
5.4 Conclusions 96

Chapter 6. *Adaptive Significance of Circannual Rhythms* 99

6.1 General Advantages of Circannual Rhythms . . . 99
   6.1.1 Improvement of Consistency of Seasonal Timing 99
      6.1.1.1 Hibernating Mammals 100
      6.1.1.2 Migratory Birds 101
   6.1.2 Timing of Seasonal Activities in Unpredictable or Constant Environments 105
6.2 Specific Functions of Circannual Timing Mechanisms 106
   6.2.1 Timing and Adaptive Programming of Seasonal Activities in Hibernating Mammals 106
   6.2.2 Timing and Adaptive Programming of Seasonal Activities in Migratory Birds 108
      6.2.2.1 Onset of Vernal Migration 108
      6.2.2.2 Onset of Autumnal Migration 110
      6.2.2.3 Pattern of Autumnal Migration . . . 114
      6.2.2.4 Duration of Autumnal Migratory Activity – A Factor Determining Migratory Distance? 116
      6.2.2.5 Migratory Direction 125
      6.2.2.6 Photoperiodic Modification of the Program 125
6.3 Conclusions and Perspective 128

Appendix. *General Oscillator Model and Terminology* . 131

References 135

Systematic Index 151

Subject Index 153

Chapter 1

Introduction

1.1 The Phenomenon of Annual Rhythmicity

Most long-lived organisms that inhabit seasonally changing environments have evolved control systems that adapt them to annual fluctuations of biologically significant factors. This is seen in the concentration of many biological activities at times of the year when they are most likely to be successful. In temperate zones, where conditions of late spring and summer are utilized by most animals for rearing the offspring, reproductive activities are timed accordingly. During winter, many species adapt to severe environmental conditions by reducing metabolic activity into a state of dormancy, diapause or hibernation; others avoid unfavorable winter conditions by leaving their home areas in late summer or autumn to migrate to places with a more propitious climate. Processes like the molt of skins, furs, and plumages are often timed to occur between the reproductive season in spring and summer, and the periods of migration or quiescence in autumn or winter (for reviews see, e.g., Lack 1950; Aschoff 1955; Immelmann 1963a, b, 1967, 1971, 1973; Murton and Westwood 1977; Farner and Follett 1979; Farner and Gwinner 1980). In the course of a year these and other activities constitute annual "phase-maps" that specify the time at which different processes occur at the level of the species, the population, or the individual.

Annual biological rhythms are most clearly expressed in organisms inhabiting temperate and arctic zones, where seasonal differences in environmental conditions are most pronounced; but similar rhythms are also present in many tropical plants and animals. Truly arrhythmic patterns in biological activities are scarce and have for instance been reported for organisms inhabiting arid regions like the central Australian deserts, where rainfall is highly unpredictable (for reviews see, e.g., Baker 1938a, b; Keast and Marshall 1954; Aschoff 1955; Keast 1959; Miller 1960; Immelmann 1963a, b, 1967, 1971).

1.2 Ultimate and Proximate Factors in the Control of Annual Rhythms

1.2.1 Ultimate Factors

It is probably safe to assume that natural selection favors such individuals that carry out their various activities during times when environmental conditions are most favorable for survival of both parents and offspring. Those environmental variables that, in the course of evolution, exert selection pressure to restrict an activity to a particular time of the year have been called "ultimate causes" by Baker (1938a) and "ultimate factors" by Thomson (1950).

Ultimate factors differ greatly, depending on species and seasonal activity, and even a particular seasonal activity may be timed relative to more than one ultimate fator. A large food supply, sufficient for raising the offspring and covering the energetic costs of the parents, is assumed to be a major ultimate factor for reproduction (e.g., Lack 1950). Indeed, in all species studied so far the critical phases of the reproductive season coincide with the maximal abundance of preferred food items. This coincidence is most conspicuous among food specialists in extreme environments. The red-backed sandpiper (*Calidris alpina*) in Alaska provides an example. Its young normally hatch when adult dipterans, the major food items of the chicks, emerge. If the young hatch before or after that time, they suffer high mortality (Holmes 1966). The role of food as an important ultimate factor in temporal control of reproduction is also indicated by those few temperate-zone species of birds that do not breed in spring. Such species have evolved a dependency on trophic resources that are available only in other seasons. For example, Eleonora's falcon (*Falco eleonorae*), which inhabits small Mediterranean islands, breeds mainly in August and September, when it feeds its young almost exclusively on small passerine birds that cross the Mediterranean in large numbers on their autumnal migration (Walter 1968).

Although food can be considered the dominating ultimate factor in the timing of reproduction, it is not the only one. There are avian species for which the availability of appropriate nest sites is an ultimate factor in timing of reproduction. Examples are found in tropical savannah species that begin nest building late in the wet season, after the grass has grown long enough to provide cover. Among tropical birds that nest in river banks (like swallows, bee-eaters, kingfishers) there are species that breed late in the dry season, when water levels are low, hence providing more potential nesting sites. In arctic species, breeding is partly timed relative to the melting of snow and ice. Lastly, other ultimate factors related to predation and competition have also been suggested (see Immelmann 1971 for a review).

1.2.2 Proximate Factors

Because they require relatively long periods of development, most seasonal functions cannot be initiated instantaneously when ultimate factors become optimal. This is especially true for reproduction, for which the main ultimate factor, food, exerts its action toward the end of the reproductive season when the offspring require an increasing amount of food. To assure this coincidence of abundant trophic resources and maximum food requirement, the preceding processes of gonadal development, courtship, nest building, and gestation or incubation must be initiated long in advance, often at times when conditions are still far from optimal. Similarly, the migrations of many organisms require thorough preparation, as their metabolism has to be adapted to the energetic demands of migration. The necessity for these preparatory periods makes it difficult or impossible to time the seasonal functions directly by the ultimate factors (or rather some critical values thereof) because they occur too late. The fitness of an individual is doubtless enhanced if it is capable of obtaining predictive information concerning environmental conditions to come. Thus mechanisms have evolved for use of reliable, predictive information so that the individual is prepared for the times at which the ultimate factors are optimal. Baker (1938a) called cues of this nature "proximate causes", whereas Thomson (1950) later designated them "proximate factors". They control annual cycles by regulating physiological processes, as opposed to ultimate factors, that exert their effects during evolution by changing gene frequencies.

The extent to which an organism is able to select proximate factors differing from the ultimate factors depends on its ecological requirements, the properties of its environment and the activity to be timed. In general, organisms inhabiting relatively unpredictable environments rely on proximate factors that are identical with or closely related to the ultimate factors in a temporal sense. In contrast, organisms inhabiting highly predictable environments can use proximate factors that are qualitatively different and temporarily separated from the ultimate factors. Prime examples of unpredictable environments are west and central Australian deserts, where life depends to a high degree on rainfall. Since rains in these areas occur irregularly and the favorable conditions following rainfalls are usually short, organisms must respond quickly to take advantage of them. Many species of bird of these areas have evolved highly sophisticated strategies to cope with this difficult situation (e.g., Keast and Marshall 1954; Keast 1959; Marshall 1959; Immelmann 1963a, b; Serventy 1971). In contrast to most other avian species which have an extended period of gonadal quiescence, these birds maintain their reproductive systems permanently at a relatively high state of development. As a result they are able to respond almost immediately to stimuli related to rainfall. This is true of the zebra finch (*Taeniopygia guttata*), whose gonads remain partially developed throughout the year. Intense courtship and nest building begin with the first showers, regardless of the length of the preceding drought period (Immelmann 1963a, b). Egg laying may begin as early as 2 weeks after the first rainfall. Obviously, stimuli related directly to the rain provide the proximate factors. It has been proposed that the sight or the noise of falling rain may be of significance, as also changes in the osmotic state of the birds, but experimental evidence is lacking. Recently, Priedkalns and Bennett (1978) and Priedkalns et al. (1984) demonstrated experimentally that testicular growth could be initiated by increasing relative humidity, a finding that is consistent with field observations indicating that nest building may begin even before it actually begins to rain.

In temperate and arctic regions, many of the environmental factors relevant for an organism vary in a rather regular annual fashion with characteristic and relatively constant mutual temporal relationships. Consequently, almost any environmental variable contains predictive information about the others and hence can be exploited as a proximate cue. Obviously, ideal proximate factors should show regular and consistent annual changes. Ambient temperature, which fulfills this criterion to a certain extent, is indeed used as a proximate factor in many mid- and high-latitude species, especially poikilotherms (e.g., Danilevskii 1965; Hoar

1969; Lofts 1974; Saunders 1976). An even more reliable cue, however, is photoperiod, the light fraction of the 24-h day, which among all obvious environmental factors shows the least year to year variability. It is understandable therefore that in the evolution of control systems of numerous temperate-zone plants and animals the annual cycle in photoperiod has become the major source of predictive environmental information in the control of a variety of seasonal activities. In fact, "there is no other environmental factor in any climatic region which is of comparable importance for the immediate control of annual cycles" (Immelmann 1973). The functions that are controlled by photoperiod include essentially all classes of activity known to show seasonal variations (for reviews see e.g., Farner et al. 1973; Vince-Prue 1975; Follett and Davies 1975; Murton and Westwood 1977; Turek 1978; Farner and Follett 1979; Farner and Gwinner 1980; Follett and Robinson 1980; Hoffmann 1981; Saunders 1981).

1.3 Circannual Rhythms

The evolution of photoperiodic response systems in organisms of temperate and higher latitudes had long been thought to represent the final stage in the process of becoming independent from the direct control of seasonal activities by ultimate factors. Only recently has it become clear that in some species this process of emancipation has developed further in that a great deal of the seasonal timing machinery has been incorporated into the endogenous organization of the organism. This is most conspicuously so in species whose seasonal cycles are preprogrammed into an endogenous circannual rhythmicity. Normally synchronized with the natural year, circannual rhythms have taken over a great deal of the task of timing seasonal activities both relative to the external world and relative to each other. Like the other three classes of biological rhythm that provide adaptations to environmental cycles – circadian, circatidal, and circalunar rhythms – circannual rhythms function as clocks to adjust the physiology and behavior of an organism to the periodically changing conditions of the world in which it lives. The question of how this is achieved is the subject of this book. (For previous reviews see, e.g., Berthold 1974c; Pengelley 1974; Davis 1976; Mrosovsky 1978; Gwinner 1981c d; Canguilhem 1985).

1.4 Hierarchical Organization of Proximate Factors

In the preceding sections it was attempted to arrange proximate control mechanisms of seasonal biological phenomena according to their degree of remoteness from the ultimate factors. To simplify matters, emphasis was placed on those proximate factors that provide the dominating timing cues. However, as with ultimate factors, more than one proximate factor is usually involved in the control of a particular seasonal activity. Quite often we find a situation in which the action of a primary proximate factor is enhanced by that of a variety of secondary stimulatory or inhibitory cues. This holds especially true for cases in which the dominating proximate factors are very different and temporally separated from the ultimate factors, and/or for functions that are strongly controlled by a circannual rhythmicity. Here, the dominating proximate causes are often only responsible for the crude preparation of the organism to perform a particular seasonal activity at about the appropriate time. The fine tuning is then achieved by stimuli that are more closely related to the ultimate factors (reviews: Immelmann 1971; Farner and Follett 1979; Farner and Wingfield 1980; Wingfield and Farner 1980; Farner 1985). Examples are found among those migratory birds in which a circannual rhythmicity, synchronized by the annual photoperiodic cycle, determines when the animals come into general migratory condition, whereas the stimuli that actually release (or inhibit) migration are provided by the weather situation or food conditions (Berthold 1975b; Gwinner 1986). Similarly, in many female birds breeding at higher latitudes, photoperiodic stimulation leads only to a partial development of the ovary, whereas vitellogenesis and the final follicular development require supplementary information from the environment, for example, stimulation by a territorial male or the availability of adequate food (Immelmann 1973; Murton and Westwood 1977; Farner and Follett 1979; Farner and Wingfield 1980; Wingfield and Farner 1980).

1.5 Early Suggestions of Circannual Rhythms

1.5.1 Mammalian Hibernation and Avian Migration

The participation of endogenous timing mechanisms in the control of annual rhythms was proposed long ago, particularly for organisms that spend part of the year in a constant or unpredictable environment, but still must organize their activities according to an annual rhythmicity. Some hibernating mammals, for instance, spend many months each year in deep torpor underground in constant darkness and at constant temperatures. It was suspected earlier on that their emergence from hibernation in spring was not a simple response to an environmental stimulus, but due to the action of endogenous timing factors (Berthold 1837; Dubois 1896; Kayser 1940; Lyman 1948, 1954). Pengelley and Fisher (1957, 1963) later confirmed this conjecture. Their investigations on golden-mantled ground squirrels (*Spermophilus lateralis*) provided the first clear demonstration of a circannual rhythmicity in any organism (Chap. 2.1.1).

Many migratory birds spend the winter in equatorial regions where external information about season is poor or missing. Nevertheless, they begin homeward migration at a precise time in early spring. The idea that such birds rely on endogenous factors for the timing of migrations can be traced back to von Pernau (1702), who noted that migratory birds were being "driven at the proper time by a hidden drive" (translated). Similar statements have been made by other early investigators of bird migration like Naumann (1822), Brehm (1828), and von Homeyer (1881). Later, Rowan (1926) was very explicit in stating that the photoperiodic control mechanisms, which he himself discovered for birds, could not explain the seasonal cycles of long-distance migrants: "Those species that breed in the northern hemisphere and winter on the equator or cross it and winter in the southern hemisphere make necessary the assumption that there is another and internal factor involved, a physiological rhythm." This hypothesis was generally accepted by many other authors (e.g., Chapin 1932; Marshall 1951, 1959, 1960a, b; Aschoff 1955; Merkel 1956, 1963), but it was not rigorously tested or experimentally verified before the late 1960's (Gwinner 1967, 1968a).

1.5.2 Yearly Breeding Cycles in Tropical and Temperate-Zone Animals

Species living permanently in tropical areas were also thought to be equipped with endogenous annual rhythms (e.g., Moreau 1931; Chapin 1932; Baker and Baker 1934/1936; Baker and Ranson 1938; Moreau et al. 1947; Marshall 1959, 1960a, b; Marshall and Serventy 1959); furthermore there was even some early evidence suggesting that the same might be true of some temperate-zone species (see Aschoff 1955 for a review of the early literature). Displacement experiments with animals across the equator led Baker (1938b) to conclude that "there is also an internal rhythm in reproduction which may be so strong as to cause specimens of southern hemisphere birds, imported into the northern hemisphere, to continue breeding at the same time as the others of their species in the south." In another paper, Baker and Baker (1934/1936) even discussed the analogy of this postulated internal rhythm with a clock. Based on studies on the control of gonadal cycles in both tropical and temperate zone birds, Marshall (1951) expressed the belief "that the internal gonadal rhythm is the most important single factor in the timing of the breeding seasons and the migration that is part of them." Similarly Blanchard (1941) concluded from the investigations of the reproductive cycles of white-crowned sparrows (*Zonotrichia leucophrys*): "It seems nearest to the truth, then, to think of the gonad cycle as the expression of an inherent annual rhythm... which may be modified in part by environmental conditions but is by no means entirely dependent upon them for its beginning or its general subsequent course."

Several investigators of tropical species considered the "internal rhythm" as a mechanism that could fully explain the seasonal nature of reproduction, migration, and molt in areas where environmental factors show only weak annual variations. That this explanation could possibly not be sufficient was early recognized by Baker (1938b): "Internal rhythm can never account wholly for the timing of breeding seasons, for it would get out of step with the sun in the course of the ages, but it is likely that it plays its part in making a species quick to respond to the external factors." This statement of Baker is an early intuition of the phenomenon of free-running rhythms that became a crucial issue in the development of the field.

1.5.3 Ten-Month Breeding Cycles in Tropical Birds

In a few avian species the breeding seasons do indeed "get out of step with the sun" in that they occur at intervals of about 10 months rather than 12. The most famous example is that of the sooty tern (*Sterna fuscata*), which breeds on the Ascension Islands, close to the equator. These birds begin incubation on the average every 9.7 months (Chapin 1954; Chapin and Wing 1959). Similar cycles occur in several other species of tropical seabird (Immelmann 1971, for a review). In addition, Fogden (1972) found approximately 10-month breeding and molting cycles in two passerine species in Borneo, a babbler (*Stachyris erythroptera*) and a sunbird (*Arachnothera longirostris*). These cases are frequently cited as evidence for an internal annual rhythmicity, but the possibility also exists that the rhythms are controlled by the moon, as their period is indistinguishable from 10 lunar cycles (Chapin and Wing 1959). If they were, in fact, free-running circannual rhythms, one would have to assume that individuals of a population synchronize each other by social zeitgebers.

1.5.4 Plants

The botanical literature contains many early inklings of possible endogenous annual rhythms. According to Bünning (1956), botanists and gardeners early this century discovered that plants taken from Europe to the tropics continued to show periodicities in leaf unfolding, flowering, fruiting etc. Since the periods of these rhythms deviated from 12 months, the synchrony with the natural year was no longer maintained. Even the rhythms of different individuals and of different branches of a plant became asynchronous. Unfortunately, these interesting reports have not been validated by further experimental investigations.

In many indigenous plants of the tropics, particularly the humid tropics, where seasonal variations in environmental factors are often extremely small, growth, leaf shedding, and flowering occur in a rhythmic fashion. In some cases these rhythms were reported to have periods close to but different from one year, suggesting once again endogenous mechanisms as the basis of these periodicities (Dingler 1911; Volkens 1912; Simon 1914; von Ihering 1923; Resende 1947; Koriba 1948; for reviews see Bünning 1949a, b, 1955, 1956; Remmert 1965).

Sperlich (1919) was one of the first to propose that some of the processes that occur in seeds, particularly the quiescent period that precedes germination ("seed dormancy") are of an endogenous nature. He suggested that these processes are based on the internal structure of the embryo or endosperm. "They require a certain amount of time for their course and in this way resemble a music box which can reproduce its tune in the same way anywhere and at any time" (translated). This early intuition of the involvement of a circannual rhythmicity in the germination of seeds has been supported by subsequent experimental work (see Chap. 2.1.5).

1.5.5 Development of the Oscillator Analogy

The early studies on circannual rhythms suffered from the lack of a conceptual framework and an appropriate terminology for describing the phenomena and formulating the appropriate questions. This situation changed radically when in the late 1950's and early 1960's Aschoff and Pittendrigh, working in the related field of circadian rhythm research, developed a general oscillator model, in which biological rhythms were treated as oscillators in the technical sense (e.g., Aschoff 1960; Pittendrigh 1960; cf. Aschoff 1981b for terminology). Applying this model to animal rhythmicities, Aschoff (1955) outlined the scope for future research on circannual cycles. His paper provides a landmark in the field and has stimulated a great deal of the research summarized in this book. The main features of the general oscillator model and its terminology are presented in the Appendix (for more details see, e.g., Aschoff et al. 1965; Aschoff 1980, 1981b).

1.6 Some Definitions and Delimitations

To demonstrate that a rhythmicity is endogenous, it is necessary to show that it continues in conditions that provide no external information about the period it normally assumes. In view of the fact that numerous environmental factors vary on a tidal, daily, lunar, or annual basis, it seems very difficult to show that biological rhythms with one of these periods are truly endogenous. Whatever precautions an experimenter may take to isolate an organism from these environmental cycles, a critic could always still maintain that some relevant (and perhaps even entirely unknown) factor might have penetrated the experimental chamber and caused the observed rhythmicity. Indeed, in the related field of circadian rhythm research, scientists were vigorously struggling with this very problem, for instance, in discussions about the possibility that an unknown environmental "factor X" could be responsible for daily rhythmicities in "constant" conditions (e.g., Bünning 1973). Fortunately, however, annual biological rhythms, just like tidal, daily, and lunar rhythms, behave in a way that excludes this kind of objection. As will be illustrated in later chapters, the period of these rhythms is usually different from one year, if the organism is isolated from the relevant synchronizing factors. In other words, the subjective year of an organism deviates from the objective year in these conditions, because its annual "clock" runs either fast or slow. As with circadian rhythms, this deviation of the period of the biological rhythmicity from that of the corresponding environmental rhythmicity provides the strongest evidence for the endogenous nature of rhythmicity, justifying the use of the term "circannual" for this class of biological rhythm.

If, conversely, the period of an annual cycle is very close to one year, the suspicion exists that some seasonal environmental cues may have caused rhythmicity. This reservation is valid for a host of studies in which annual variations in a variety of different functions have been demonstrated in laboratory animals, especially mice and rats (e.g., von Mayersbach 1978; Wong et al. 1983). Since the present monograph concentrates on truly endogenous annual rhythms, these data will not be further considered, despite the fact that they raise some interesting questions about the subtle environmental cue(s) that might be involved in generating these annual cycles.

There has been some discussion in the literature as to which conditions should be considered sufficiently constant to justify the conclusion that an annual rhythmicity is truly endogenous. Specifically, the question has been raised of whether keeping an organism in a 24-h light-dark cycle (LD) with an invariable photofraction – rather than in continuous light or darkness – fulfills the criteria of constant conditions (King 1968; Hamner 1971; Sansum and King 1976; Farner 1985). As stated before, the demonstration of the endogenous nature of a rhythm requires that it persists in conditions that provide no information about its period. If one accepts this criterion, a 24-h light-dark cycle with an unchanging photofraction is a constant condition for an annual rhythmicity, as it contains no information about the duration of a year. If one did not accept this criterion, and instead demanded that the organism be isolated from all periodic input (including the daily) the demonstration of an endogenous annual rhythmicity would be a priori impossible, since there is no way to isolate an organism from all conceivable periodic stimuli.

Accepting that an unchanging 24-h light-dark cycle should be considered a constant condition for an annual rhythmicity, does not, of course, imply that the properties of such a light-dark cycle have no effects on the circannual rhythmicity. On the contrary, it almost has to be expected, both on theoretical grounds and by analogy with the behavior of circadian rhythms under various intensities of constant light (e.g., Aschoff 1979), that the duration of photoperiod affects the angular velocity and/or the overall period of such circannual rhythms which are normally synchronized by the annual variations in photoperiod. Consequently, the problem of the dependence of circannual systems on properties of the light-dark cycles will be considered extensively in subsequent chapters (e.g., Chap. 5).

Closely related to the apparent problem of constant conditions is the question of "permissive" conditions under which circannual rhythms can be expressed. As will be shown in detail later (e.g., Chaps. 3 and 5), there is a bewildering diversity of environmental conditions, especially those of photoperiod, under which certain animals do or do not exhibit circannual cycles. The analysis of both the physiological basis and the functional meanings of these limitations is a major challenge in circannual rhythm research (Chaps. 5 and 6; see also Gwinner and Dittami 1986). The existence of conditions in which circannual rhythms are not present, should, however, not prevent us from formally designating them as endogenous or circannual, as long as they can be observed in others. A similar phenomenon of a rather limited range of environmental conditions, permissive for the expression of rhythmicity, is also well known (but hardly investigated) among circadian rhythms (e.g., Aschoff 1979).

Chapter 2

Evidence for Circannual Rhythms

2.1 Typical Cases

2.1.1 Mammals

Although a considerable amount of evidence suggesting endogenous annual clocks in plants and animals was accumulated during the first half of this century, their existence was not demonstrated before the early 1960's. The most compelling evidence then was provided by the work of Pengelley and Fisher (1957, 1963) with golden-mantled ground squirrels (*Spermophilus lateralis*). Figure 2.1 shows the seasonal changes in body weight and food consumption, as well as the occurrence of hibernation, in two ground squirrels maintained for about 2 years under a constant LD 12:12 and constant temperature conditions. Both animals continued to hibernate about once a year, and each period of hibernation was preceded by a dramatic increase in body weight and food consumption. The general pattern of these seasonal functions was rather similar to that in free-living squirrels with an obvious difference that can be seen particularly clearly in the squirrel maintained in 0 °C. In the first year of the experiment this animal entered hibernation in late October, but in the second the onset of hibernation had shifted forward to mid-August and in the third to early April. In other words, the period measured between onsets of hibernation was shorter than 12 months.

The existence of circannual rhythms in golden-mantled ground squirrels has been confirmed and extended in later studies by Pengelley and his coworkers as well as by other authors (see Tables 2.1 and 2.2). Figure 2.2 summarizes the results of a more recent experiment (Pengelley et al. 1976a). It shows the occurrence of hibernation in five groups of squirrels kept for 47 months in constant darkness (DD), constant light (LL) or in a constant LD 12:12; some of the animals were blind, the others sighted. In all individuals hibernation continued to occur about once a year throughout the experiment. Again, the period of the rhythm deviated from exactly 12 months. This can be seen easier in the diagram on the right. Here, the symbols connected by lines indicate the mean dates at which the animals of each experimental group commenced hibernation in successive years. Most animals entered hibernation earlier each year than in the previous year. The period of the annual hibernation cycle was shorter than 12 months, irrespective of the lighting conditions and irrespective of whether the animals were blind or sighted.

Circannual rhythms of hibernation and related functions persisting for at least two cycles under seasonally constant conditions have also been documented for several other hibernating rodents including at least four other species of ground

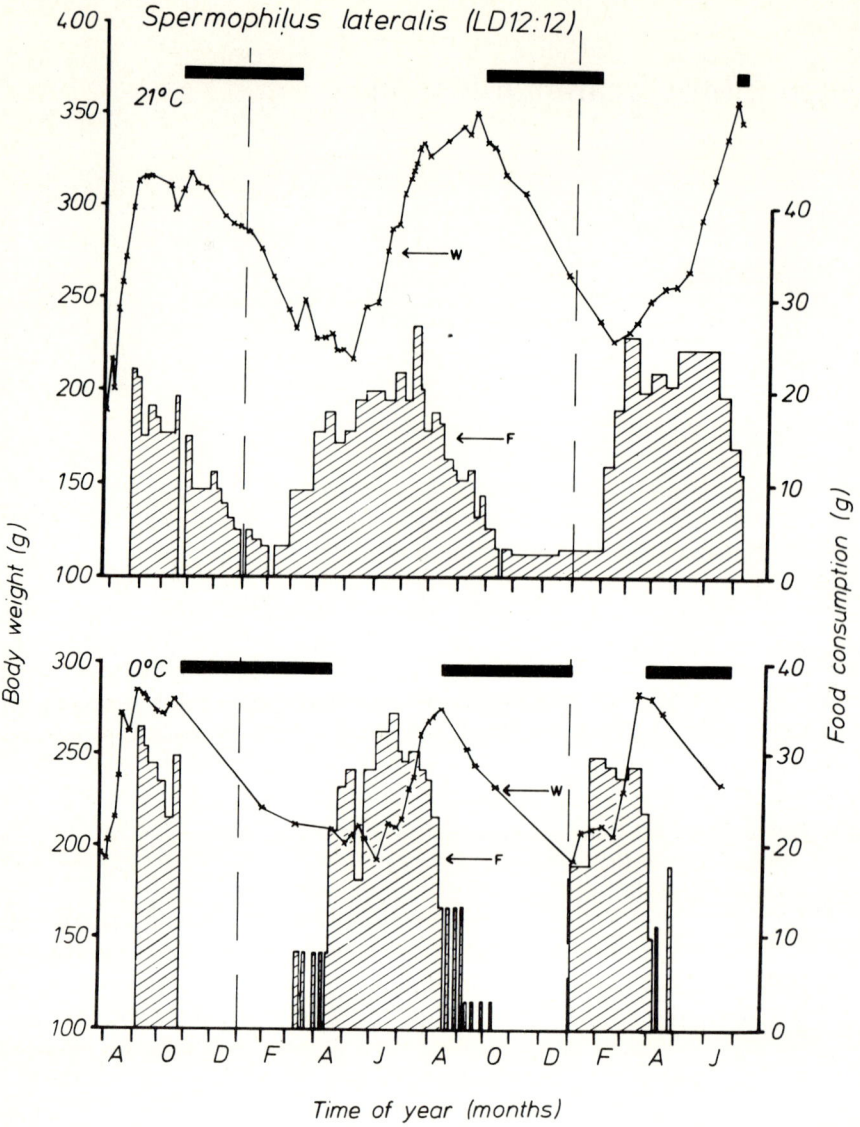

Fig. 2.1. Circannual rhythms in body weight (*W, curves*), daily food consumption (*F, hatched histogram*) and hibernation (*black bars*) in two golden-mantled ground squirrels (*Spermophilus lateralis*) held for 23 months in constant LD 12:12 and under constant temperatures of 21 ° oder 0 °C. (After Pengelley and Fisher 1963)

Table 2.1. Summary of investigations providing strong evidence for circannual rhythms. In the studies listed in this table at least two cycles were measured in constant conditions with periods clearly deviating from 12 months

	Species	Light conditions	Temperature °C	Maximal time in constant conditions (months)	Functions showing periodicity	Maximal number of complete cycles	Estimated period (months)[a]	Ref.
Plants	*Lemna minor*	LL	25	43	Dry matter production	2	(11 and 14)	Bornkamm (1966)
Coelenterates	*Campanularia flexuosa*	DD	10, 17	39	Growth and longevity of hydrants	3	12–13	Brock (1975a–c)
Molluscs	*Limax flavus*	LD 11:13	10, 20	36	Oviposition	2	11	Segal (1960)
Arthropods	*Orconectes pellucidus*	DD	13	30	Gonadal size, molt	2	11–13	Jegla and Poulson (1970)
	Anthrenus verbasci	DD	15, 17.5, 20, 22.5, 25	35	Diapause	2	10–11	Blake (1959a, b)
Fish	*Catostomus commersoni*	LD 12:12	15	19	Behavioral thermoregulation	2	10	Kavaliers (1982)
Reptiles	*Sceloporus virgatus*	LD 9:15	16.5	21	Activity	2	7–10	Stebbins (1963)
Birds	*Parus cristatus*	LD 10:14	20	22	Molt	2	10	Berthold (1973a, 1980)
	Phylloscopus trochilus	LD 12:12	21	28	Molt, zugunruhe	3	9, 11.8	Gwinner (1967, 1968a, 1971a)
	Sylvia borin *S. atricapilla*	LD 10:14 LD 12:12 LD 16: 8	20	34	Body weight, molt, zugunruhe, gonadal size	3	10.6[b] 10.5[b]	Berthold et al. (1971, 1972a, b)
	Sylvia borin	LD 11:11	20	30	Size of testes or follicles	3	(10)	Gwinner and Dorka (1976)
	Sylvia borin *S. atricapilla*	LD 10:14	20	96	Molt	9	9.4; 9.7	Berthold (1978)
	Sylvia borin	LD 11:11 LD 12:12 LD 13:13	20	30	Testicular size	3	10–13	Gwinner (1981a)

Table 2.1 (continued)

	Species	Light conditions	Temperature °C	Maximal time in constant conditions (months)	Functions showing periodicity	Maximal number of complete cycles	Estimated period (months)[a]	Ref.
	Sylvia undata	LD 10:14	20	19	Molt, zugunruhe	2	11.1[c]	Berthold (1974a, b)
	S. melanocephala						10.5[c]	
	S. sarda						10.4[c]	
	Sylvia cantillans	LD 10:14	20	19	Body weight, molt, zugunruhe	2	11.3[c]	Berthold (1974a, b)
	Ficedula albicollis	LD 13:11	20	31	Body weight, molt, zugunruhe	2	(13)	Gwinner and Schwabl-Benzinger (1982)
	Ficedula hypoleuca	LD 12:12	20	30	Body weight, molt	2	(13)	
	Sturnus vulgaris	LD 12:12	22–28	28	Testicular size	2	(10)	Schwab (1971)
	Sturnus vulgaris	LD 11:11	20	40	Molt, testicular size	3	(11)	Gwinner (1977a)
		LD 12:12						
	Sturnus vulgaris	LD 12:12	20	27	Molt, testicular size	2	8–9	Gwinner et al. (1980)
	Sturnus vulgaris	LD 11:11	20	43	Molt, testicular size	3	9–15	Gwinner (1981a)
		LD 12:12						
		LD 13:13						
	Fringilla coelebs	LD 12:12	20–22	23	Molt, gonadal activity, fat deposition	2	8–13	Dolnik (1974)
		LD 20:4						
	Loxia curvirostra	LD 12:12	20	28	Body weight	2	9.6	Berthold (1977b)
	Zonotrichia leucophrys	LD 12:12	20	46	Testicular size	4	(11)	Farner et al. (1980)
	Lonchura punctulata	LL	27	32	Testicular size	2	10	Chandola et al. (1982)
		LD 12:12						
	Lonchura punctulata	LL	27	30	Body weight, food intake, testicular size	2	10	Bhatt and Chandola (1985)
Mammals	*Macaca mulatta*	LD 14:10	22	36	Plasma androgens	3	(13)	Michael and Bonsall (1977)
	Antrozous pallidus	LD 14:10	23	36	Body weight	3	9–10	Beasley et al. (1984)
		LD 10:14						

Species	Light			Parameters			Reference
Spermophilus lateralis	LD 12:12	0,21	23	Hibernation, body weight, food consumption	2	9–11	Pengelley and Fisher (1957, 1963)
Spermophilus lateralis	LD 12:12	3, 12	15	Hibernation	1	10–11	Pengelley and Kelly (1966)
Spermophilus lateralis	LD 12:12	3, 12	47	Hibernation, body weight	5	10–12	Pengelley and Asmundson (1969)
Spermophilus lateralis	LD 12:12 / LD 20: 4	3	40	Hibernation	3	10–12	Pengelley and Asmundson (1970)
Spermophilus lateralis	LD 12:12	3	38	Hibernation	3	10–12	Pengelley and Asmundson (1975)
Spermophilus lateralis	DD, LL	3	47	Hibernation	4	10–12	Pengelley et al. (1976a)
Spermophilus lateralis	LD 12:12	5, 34	36	Hibernation, body weight	2 (3)	11,3 / 10,9	Pengelley et al. (1978)
Spermophilus lateralis	LD 12:12	5	55	Hibernation, body weight	5	10–12	Pengelley et al. (1979)
Spermophilus lateralis	LD 12:12	16	33	Hibernation, body weight, reproductive condition	2	10	Heller and Poulson (1970)
Spermophilus lateralis	LD 12:12	0, 18	56	Hibernation	4	11	Scott and Fisher (1970)
Spermophilus lateralis	LD 12:12	12.5, 22	31	Hibernation, body weight	4	9–14	Mrosovsky (1975)
Spermophilus lateralis	LD 12:12	9.5, 21	34	Hibernation, body weight	2	12.2 / 10.4	Mrosovsky (1980b)
Spermophilus lateralis	LD 10:14	23	29	Body weight, estrous	3	11	Zucker (1985b)
Spermophilus lateralis	LD 14:10	23	36	Body weight	3	10–12	Dark et al. (1985)
Spermophilus beldingi	LD 12:12	16	25	Hibernation, reproductive condition	2	15	Heller and Poulson (1970)
Spermophilus tridecemlineatus	LD 12:12	21	22	Molt, body weight	2	8–11	Joy and Mrosovsky (1982)
Spermophilus tridecemlineatus	LD 12:12	21	32	Molt, body weight	2	10–12	Joy and Mrosovsky (1985)
Spermophilus beecheyi	LD 12:12 / LL (2 lux) / LL (100–500 lux)	20 / 22	68 / 70	Body weight, torpor / Body weight, molt	5 / 7	(10) / 10–13	Davis and Swade (1983)
Spermophilus beecheyi	LD 12:12	20	70	Body weight, nest building	5	10–12	Davis (1984)
Ammospermophilus leucurus	LD 12:12	23	36	Body weight, water consumption	3	11	Pengelley et al. (1976b)

Table 2.1 (continued)

Species	Light conditions	Temperature °C	Maximal time in constant conditions (months)	Functions showing periodicity	Maximal number of complete cycles	Estimated period (months)[a]	Ref.
Eutamias speciosus *E. alpinus* *E. amoenus* *E. minimus*	LD 12:12	5, 16	33	Hibernation, reproductive condition	2	9–12	Heller and Poulson (1970)
Tamias striatus	LD 12:12 (blinded)	23–24	78	Locomotor activity, food and water intake, body weight	6	11	Richter (1978)
Marmota monax	LD 16: 8	6, 20	28	Body weight	2	(10–12)	Davis (1967)
Perognathus longimembris	LD 12:12	16	24	Surface activity	2	(10)	French (1977)
Glis glis	self-selected	12	35	Hibernation	2	12–13	Fischer et al. (1975), Butschke (1977)
Zapus princeps	DD	7	24	Hibernation	2	11	Cranford (1978)
Cervus nippon	LD 8:16 LD 16: 8 LL	Not given	42	Antler replacement	4	9–10	Goss (1969a, b)
Capra domestica	LD 16: 8	14	30	Milk production, udder volume	2	(10)	Linzell (1973)
Ovis aries	LD 6:18 LD 12:12 LD 18: 6 LL	Not given	31	Breeding activity	2	9–11	Ducker et al. (1973)
Ovis aries	LD 16: 8 LD 8:16	Variable	38	Testicular size, plasma prolactin	2	8–10	Howles et al. (1982)

[a] Numbers in paranthesis represent rough estimates.
[b] Mean values based on data on body weight, molt and zugunruhe of 25 *Sylvia borin* and 22 *S. atricapilla*.
[c] Mean values based on data on molt of at least six individuals of each species.

Table 2.2. Summary of investigations providing weak evidence for circannual rhythms. In the studies listed in this table, either only one cycle was measured in constant conditions, or, if more cycles were measured, their period was indistinguishable from 12 months

	Species	Light conditions	Temperature °C	Maximal time in constant conditions (months)	Functions showing periodicity	Maximal number of complete cycles	Estimated period (months)[a]	Ref.
Plants	*Ankistrodesmus braunii*	LL	23	24	Nitrate reduction	2	12	Kessler and Cygan (1963)
	Lemna minor	LL	20	16	Growth, osmotic value, plasmolyse	1	12	Pirson and Göllner (1953)
	Spirodela polyrrhiza	LL	20	18	Formation of perennial organs	1	12	Henssen (1954)
	Seeds of *Digitalis lutea* *Hypericum perforatum* *Potentilla molissima* *Gratiola officinalis* *Chrysanthemum corymbosum* etc.	DD	22, 3, 13 20, 35, 45, 45	20	Capacity to germinate, heat resistance, water content, catalase activity	1	(12)	Bünning (1949b) Bünning and Müssle (1951) Bünning and Bauer (1952)
	Seeds of *Dactylis glomerata*	DD	5, 25	25	Germination	1, (2)	12	Kummerow (1963)
	Phaseolus vulgaris	DD	21	12	Photoactivity	1	(12)	Spruyt et al. (1983)
Molluscs	*Helix aspera*	LD 12:12	17.5	14	Activity, food consumption, reproductive activity	1	(8–9)	Bailey (1981)
Fish	*Salmo trutta*	LD 12:12	11.5	18	Growth rate	1	(12)	Brown (1945)
	Salmo salar	LD 12:12	11	14	Body form, skin coloration	1	10	Eriksson and Lundquist (1982)
	Heteropneustes fossilis	LL DD	25	31	Ovarian weight	1	13	Sundararaj and Vasal (1973)
	Heteropneustes fossilis	LL DD LD 12:12	25	33 18	Ovarian weight	3 1	(12)	Sundararaj et al. (1982)
	Mystus vittatus	LL	Fluctuating	12	Thyroid activity	1	(10)	Singh (1968)

Table 2.2 (continued)

	Species	Light conditions	Temperature °C	Maximal time in constant conditions (months)	Functions showing periodicity	Maximal number of complete cycles	Estimated period (months)[a]	Ref.
Reptiles	*Pseudemys scripta*	LD 12:12	20	12	Feeding, blood amino acid nitrogen, blood lipoid phosphorus	1	(12)	Hutton (1960)
Birds	*Dolichonyx oryzivorus*	LL	Not given	17	Testicular width	1	9–11[b]	Hamner and Stocking (1970)
	Euplectes orix	LD 14:10	25, 15	25	Molt, testicular activity	1	9–10	Craig (1985)
	Phylloscopus collybita	LD 12:12	21	15	Molt, zugunruhe	1	(12)	Gwinner (1971a)
	Sylvia communis	LD 12:12	20	21	Body weight	1	(12)	Merkel (1963)
	Sylvia communis	LD 12:12	20	33	Body weight, molt, zugunruhe	1	(12)	Gwinner (1983)
	Sylvia borin	LD 10:14	20	9	Food preference	1	(12)	Berthold (1976a)
	Sturnus vulgaris	LL	15	20	Molt, circadian activity time, testicular size	1	(13)	Gwinner (1973)
	Sturnus vulgaris	LD 12:12	20	20	Molt, testicular size	1	14	Gwinner (1975a)
	Sturnus vulgaris	LD 12:12	16	18	Molt, testicular size	1	(12)	Gwinner and Dittami (1980)
	Fringilla montifringilla	LD 10:14	18	19	Body weight, zugunruhe	1	14–16	Pohl (1971)
	Spiza americana	LD 12:12	29	21	Body weight, molt, zugunruhe	1	(12)	Zimmerman (1966)
	Zonotrichia leucophrys	LD 8:16 LD 20:4	18	13	Body weight, molt	1	(12)	King (1968)
	Quelea quelea	LD 12:12	22	29	Testicular size	2	12	Lofts (1964)
Mammals	*Erinaceus europaeus*	LD 20:4	20	24	Plasma testosterone	2	(24)	Saboureau (1981)
	Microcebus murinus	LD 13.5:10.5	25	17	Testicular size, estrous	1	10–12	Petter-Rousseaux (1975)
	Macaca mulatta	LD 14:10	20–24	12	Aggressive and sexual behavior, plasma testosterone	1	(12)	Michael and Zumpe (1978)

Species	Light	Temp	Variable measured		Duration	Reference
Macaca mulatta	LD 12:12	22	Testicular size and function, LH, FSH, prolactin	1[c]	(12)	Wickings and Nieschlag (1980)
Vespertilio superans	LD 14:10	24	Body weight	1	10	Funakoshi and Uchida (1982)
Spermophilus lateralis	LL	22	Circadian period	1	(12)	Mrosovsky et al. (1976)
Spermophilus lateralis	LD 12:12	23	Testicular size, body weight, water consumption	1	11	Kenagy (1980)
Spermophilus lateralis	LD 12:12	22	Sleep	1	(12)	Walker et al. (1980)
Spermophilus lateralis	LD 14:10	23	Body weight, food intake	1	10	Zucker and Boshes (1982)
Spermophilus lateralis	LD 14:10	23	Plasma testosterone, plasma LH, body weight, scrotal pigmentation	1	11–12	Licht et al. (1982)
Spermophilus lateralis	LL	21	Body weight, scrotal pigmentation, locomotor activity	1	10–12	Zucker et al. (1983)
Spermophilus lateralis	LD 14:10	22	Plasma LH in castrated ♀	1	7–10	Zucker and Licht (1983a)
Spermophilus lateralis	LD 12:12	21.5	Body weight, molt	1	10–11	Joy and Mrosovsky (1982)
Spermophilus tridecemlineatus	LD 12:12	11.5	Hibernation, body weight	2	10–15	Mrosovsky and Lang (1971)
Spermophilus tridecemlineatus	LD 12:12	20	Body weight, food and water consumption, oxygen consumption	1	(12)	Armitage and Shulenberger (1972)
Spermophilus tridecemlineatus	LD 12:12	21	Body weight	1	10	Joy (1984)
Spermophilus tridecemlineatus	LD 12:12	22	Core-to-skin temperature gradient	1	(12)	Hengst and Wiebers (1984)
Ammospermophilus leucurus	LD 12:12 / LL	23 / 23	Testicular size	1	14 / 12	Kenagy (1981b)
Spermophilus mohavensis	LD 12:12	3, 12	Hibernation, body weight	1	(8)	Pengelley and Kelly (1966)
Tamias striatus	LD 12:12	0, 6, 18	Hibernation	2	(12)	Scott and Fisher (1972)
Eutamias minimus	LD 12:12	23	Body weight, testicular size	1	8–11	Kenagy (1981a)
Marmota monax	LD 16: 8	20	Food consumption	1	(12)	Fall (1971)
Marmota flaviventris	LD 12:12	24	Body weight, food consumption	1	11	Ward and Armitage (1981)

Table 2.2 (continued)

Species	Light conditions	Temperature °C	Maximal time in constant conditions (months)	Functions showing periodicity	Maximal number of complete cycles	Estimated period (months)[a]	Ref.
Marmota flaviventris	LD 9:15	22	12	Renal function, plasma osmolarity, plasma sodium, potassium	1	(12)	Zatzman and South (1981)
Glis glis	LD 12:12	5	18	Hibernation, body weight	1	11–14[d]	Mrosovsky (1977)
Eliomys quercinus	LD 12:12	12	18	Hibernation	1	(10)	Daan (1973)
Cricetus cricetus	LD 12:12 / LL / DD	7	13 / 25	Hibernation, body weight / Body weight	1 / 1	(9) / (10)	Canguilhem et al. (1973)
Cricetus cricetus	LD 12:12	23	14	Body weight, urinary volume, aldosterone excretion	1	(12)	Canguilhem and Bloch (1967)
Cricetus cricetus	LD 12:12	23	14	Sodium excretion	1	10	Haberey et al. (1967)
Cricetus cricetus	LD 12:12	15	12	Brain noradrenalin turnover	1	(12)	Kempf et al. (1978)
Mesocricetus brandti	LD 10:14 / LD 16:8	10	14	Hibernation	1	(10–12)	Hall and Goldman (1982)
Microtus agrestis	LD 12:12	18	13	Circadian phase	1	(9)	Erkinaro (1972)
Zapus hudsonius	LD 12:12 / LD 12:12	20 / 5	22.5	Hibernation, body weight	2	3.3 / 10.9	Muchlinski (1980)
Cervus nippon	LD 4.9:4.9 / LD 8:8 / LD 21:21	21	21	Testicular size, antler replacement	1	9–11	Goss (1984)
Mustela putorius	LD 14:10	Not given	22	Estrus	1	(14)	Herbert (1972)
Mustela putorius	LD 8:16	21	19	Testicular size	1	(11)	Baum and Goldfoot (1974)
Ovis aries	LD 12:12	Fluctuating	24	Sexual activity	1	11	Radford (1961)

[a] Numbers in parenthesis represent rough estimates.
[b] Only 3 out of 15 individuals.
[c] Data were collected over a 4-year period but then averaged over all 4 years.
[d]

Spermophilus lateralis

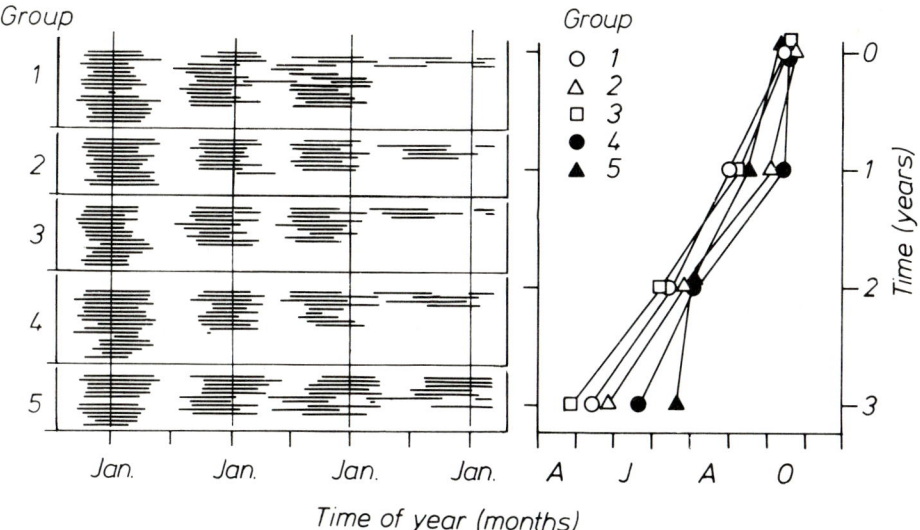

Fig. 2.2. Circannual rhythms of hibernation in five groups of golden-mantled ground squirrels (*Spermophilus lateralis*) kept for 47 months in 3 °C and under different constant photoperiods. *Group 1* DD, animals surgically blinded. *Group 2* LL (500 lx). *Group 3* LL (500 lx), animals surgically blinded. *Group 4* LL (20 lx). *Group 5* LD 12:12 (200:0 lx). *Left diagram:* Black bars arranged in horizontal rows represent hibernation periods of individual animals. *Right diagram: Symbols connected by lines* indicate mean dates at which the animals of the various groups entered hibernation in successive years of the experiment. (After Pengelley et al. 1976a)

squirrel, five species of chipmunks (*Eutamias*), the woodchuck (*Marmota monax*), the pocket mouse (*Perognathus longimembris*), the western jumping mouse (*Zapus princeps*), and the common dormouse (*Glis glis*) (cf. Table 2.1). The longest record is that of a blinded eastern chipmunk (*Tamias striatus*) kept in a constant LD 12:12 at constant temperature of 23° to 24 °C for 6.5 years (Richter 1978). Locomotor activity, food and water intake, and body weight continued to exhibit clear circannual variations without any signs of damping throughout the experiment (Fig. 2.3).

Apart from rodents, circannual rhythms have been described in several species of ungulates, for instance in sika deer (*Cervus nippon*; Goss 1969b). One experiment is summarized in Fig. 2.4. Three groups of deer were kept for up to 42 months in a constant LD 8:16, LD 16:8, or LL, and the times at which antlers were shed and regrown were recorded. Figure 2.4 indicates that the animals continued to replace their antlers at regular intervals in all the photoperiodic conditions used. The deviation of the period from 1 year is again obvious, particularly in the diagram at the right.

Examples from two domesticated species of ungulates are presented in Figs. 2.5 and 2.6. The seasonal variations in milk production and udder volume of a goat (*Capra domestica*) held over 3 years in a constant LD 16:8 are depicted in Fig. 2.5 (Linzell 1973). Figure 2.6 shows the variations in testicular volume of

21

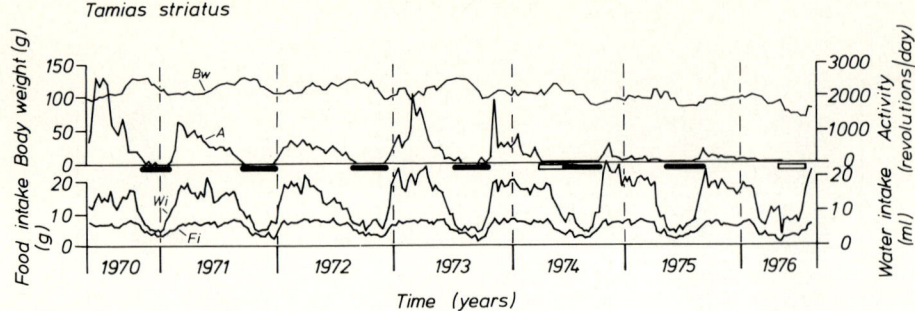

Fig. 2.3. Circannual rhythms of body weight (*BW*), wheel running activity (*A*), water intake (*Wi*) and food intake (*Fi*) in an eastern chipmunk (*Tamias striatus*) held for 6.5 years in constant LD 12:12 and under a constant temperature of 23° to 24 °C. *Horizontal black bars* indicate periods with little or no activity. (After Richter 1978)

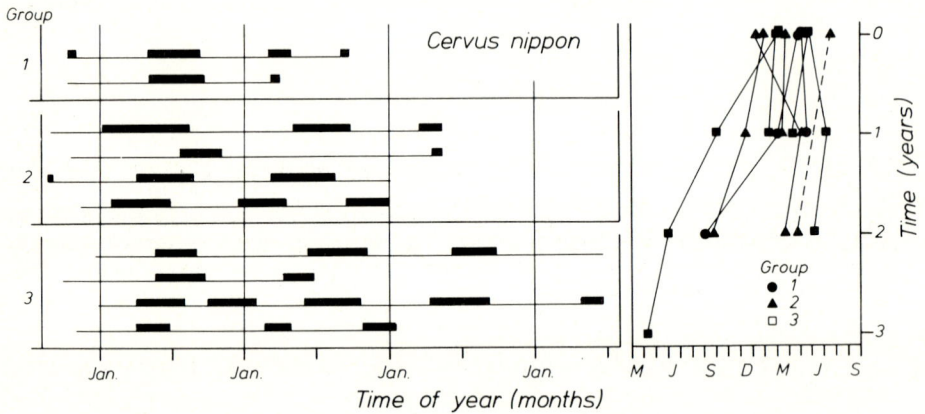

Fig. 2.4. Circannual rhythms in antler shedding and regrowth in three groups of sika deer (*Cervus nippon*) held for up to 42 months under different constant photoperiods. *Group 1* LD 8:16. *Group 2* LD 16:8. *Group 3* bright LL. *Left diagram: Black bars in horizontal rows* represent periods of antler regrowth. *Right diagram: Symbols connected by lines* indicate dates at which antlers of individual animals began to regrow in successive years of the experiment. (After Goss 1969b)

six rams (*Ovis aries*) held for 38 months in an LD 16:8 (Howles et al. 1982). In both cases the rhythms persisted throughout the experiment with periods shorter than 12 months.

Apart from the cases discussed so far, clear circannual rhythms in various functions persisting for at least two cycles with periods different from 12 months have been demonstrated in two other mammalian species, the pallid bat (*Antrozous pallidus*) and the rhesus monkey (*Maccaca mulatta*; Table 2.1). In addition there are at least ten other species of mammals for which a circannual rhythmicity has been suggested (Table 2.2). The results obtained from them are not considered entirely convincing because the animals were kept for less than 2 years under constant conditions and/or because the period of the rhythm was indistinguish-

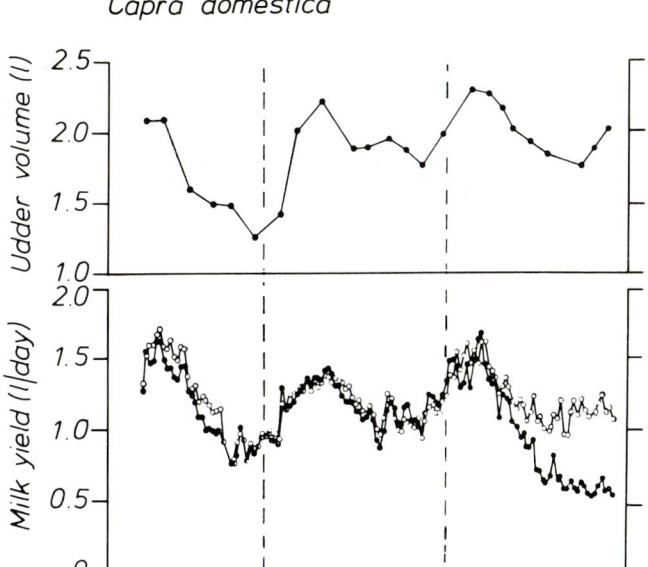

Fig. 2.5. Circannual rhythms in milk yield and udder volume of a goat (*Capra domestica*) held for 30 months under a constant LD 16:8 and under a constant temperature of 14 °C ○. Left gland, ● right gland. (After Linzell 1973)

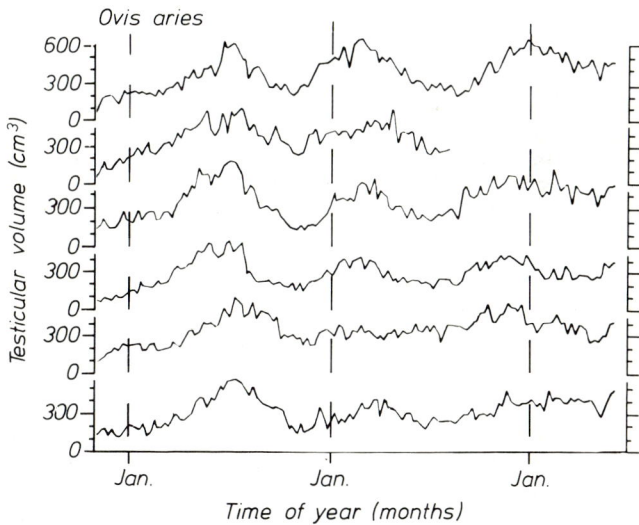

Fig. 2.6. Circannual rhythms in the testicular volume of six rams (*Ovis aries*) held for up to 38 months in an LD 16:8. (After Howles et al. 1982)

23

able from 12 months, so that the action of uncontrolled external seasonal cues cannot be definitely excluded.

2.1.2 Birds

The first evidence for circannual rhythms in birds came from investigations of a long-distance migrant, the willow warbler (*Phylloscopus trochilus*), which spends a considerable period of time each year in equatorial regions. Like many other small migrants, these birds are normally day-active but migrate at night. If kept in cages, they develop nocturnal activity, called migratory restlessness or zugunruhe, during the migratory seasons in spring and autumn, which is accompanied by an increase in body weight due to fat deposition. A complete molt is carried out twice a year, in summer and in winter. Rhythms of zugunruhe and molt were found to continue in willow warblers under a constant LD 12:12 for 28 months (Fig. 2.7). The average period was about 10 months (Gwinner 1967). The results of a more extensive study in a related species, the garden warbler (*Sylvia borin*), are summarized in Fig. 2.8. Here the rhythms of zugunruhe, molt, and body weight continued for three cycles. The average circannual period was close to 10 months, regardless of whether the birds lived in a constant LD 10:14, LD 12:12 or LD 16:8 (Berthold et al. 1972a).

The circannual molt rhythm can persist over a much longer period of time. The data shown in Fig. 2.9 are from a garden warbler and a blackcap (*Sylvia atricapilla*) which were kept in a constant LD 10:14 for over 8 years (Berthold 1978).

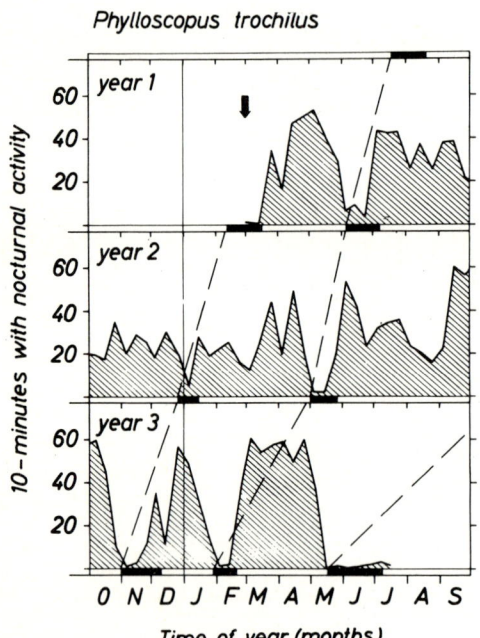

Fig. 2.7. Circannual rhythm of molt and nocturnal activity of a willow warbler (*Phylloscopus trochilus*) held for 28 months in a constant LD 12:12 and under a constant temperature of 21 °C. *Ordinate* Number of 10-min intervals with activity per night; mean values over successive 10-day periods are plotted. *Black bars* indicate periods of molt. *Dashed lines* connect successive onsets of corresponding molts. (After Gwinner 1967)

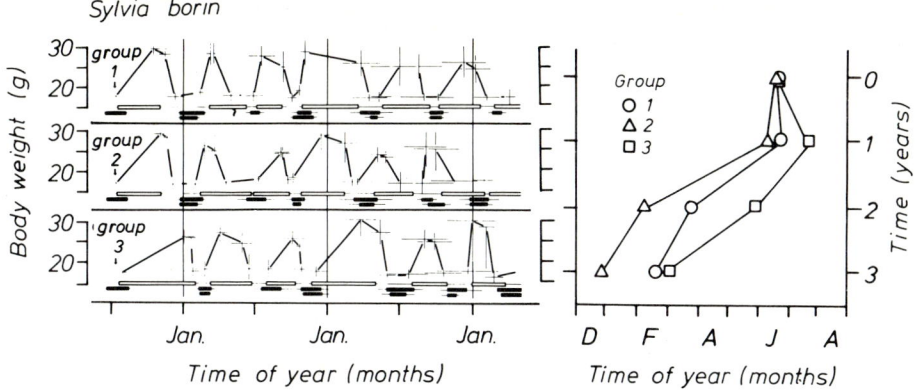

Fig. 2.8. Circannual rhythms of migratory restlessness, body weight and molt in garden warblers, *Sylvia borin,* kept for 33 months in a constant temperature of 20 °C and under three different constant photoperiods., Group 1 LD 10:14. Group 2 LD 12:12. Group 3 LD 16:8 Each group consisted of six to eight birds. *Left diagram* Curves show changes in body weight; *open bars* periods of migratory restlessness; *solid bars* periods of molt (*upper row* molt of body feathers; *lower row* molt of flight and tail feathers). *Horizontal lines at the bars* and *horizontal and vertical lines at the curve points:* standard deviations. *Right diagram:* Symbols connected by lines indicate mean dates at which the birds of the three groups initiated summer molt in successive years of the experiment. (After Berthold et al. 1972a)

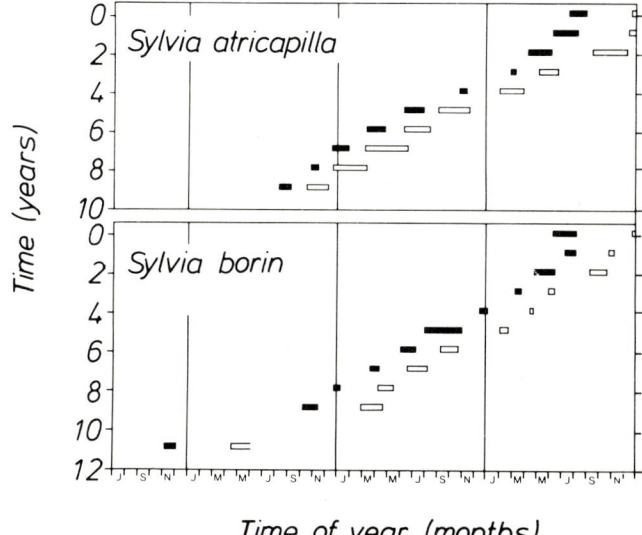

Fig. 2.9. Circannual rhythms of summer molt (*black bars*) and winter molt (*open bars*) in a blackcap (*Sylvia atricapilla*) and a garden warber (*S. borin*) kept for 8 years, respectively, under a constant LD 10:14. (After Berthold 1978)

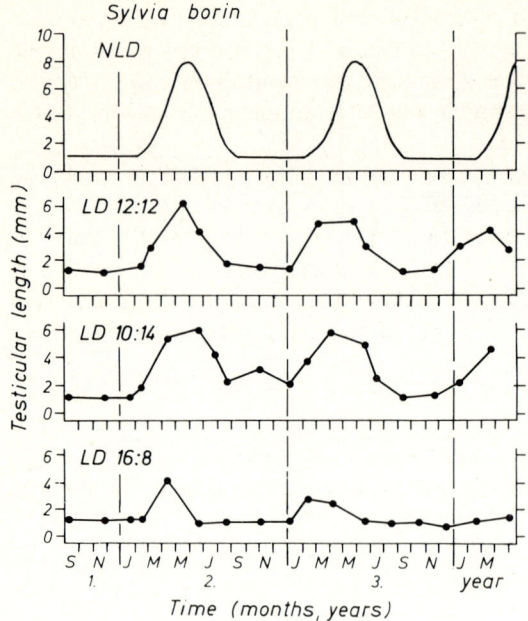

Fig. 2.10. Circannual rhythms in testicular width of three garden warblers (*Sylvia borin*) held for 32 months in a constant temperature of 20 °C and under three different constant photoperiods. The *upper diagram* is a schematic representation of the normal testicular cycle in free-living garden warblers. (After Berthold et al. 1972b)

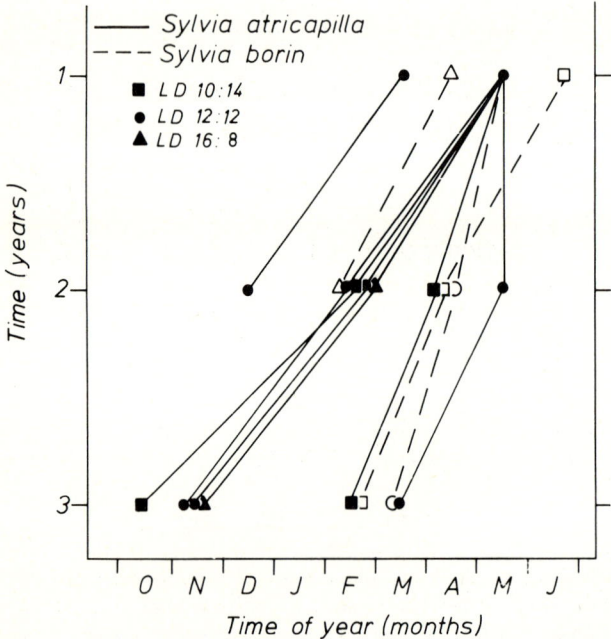

Fig. 2.11. Circannual rhythms in testicular size of three garden warblers (*Sylvia borin*) and seven blackcaps (*S. atricapilla*) kept for up to 32 months in a constant temperature of 20 °C and under the three different constant photoperiods indicated. (After Berthold et al. 1972b)

Both birds had a circannual molt rhythm with a period of about 10 months throughout the experiment. By the end of the study they had run through nine cycles within 8 calender years. Since 8 years are far beyond the normal lifespan of a free-living warbler, these results show that a circannual clock can keep functioning in these birds throughout life.

Not only molt and migratory disposition are controlled by a circannual rhythmicity, but also reproductive activity. This is illustrated in Fig. 2.10 for the rhythm of testicular size in three garden warblers held over 32 months in a constant LD 12:12, LD 10:14 or LD 16:8. These birds went through a testicular cycle about once a year similar to that of free-living conspecifics. In all three experimental birds, however, testicular growth started earlier each year than in the previous year. This circannual period of less than 12 months is more clearly seen in Fig. 2.11 where the data from these and several other birds of the same experiment are plotted by connecting the testicular maxima of consecutive years on a vertical time scale. Testicular maxima occurred progressively earlier each year, once again indicating free-running rhythms with periods of about 10 months (Berthold et al. 1972b).

Circannual testicular rhythms have been extensively studied in the European starling (*Sturnus vulgaris*). These birds usually continue to go through cyclic changes in testicular size when maintained in a constant LD 12:12 and LD 11:11. Figure 2.12 shows the changes in testicular width as well as the occurrence of molt in six selected specimens of an experiment in which birds were held for 43 months

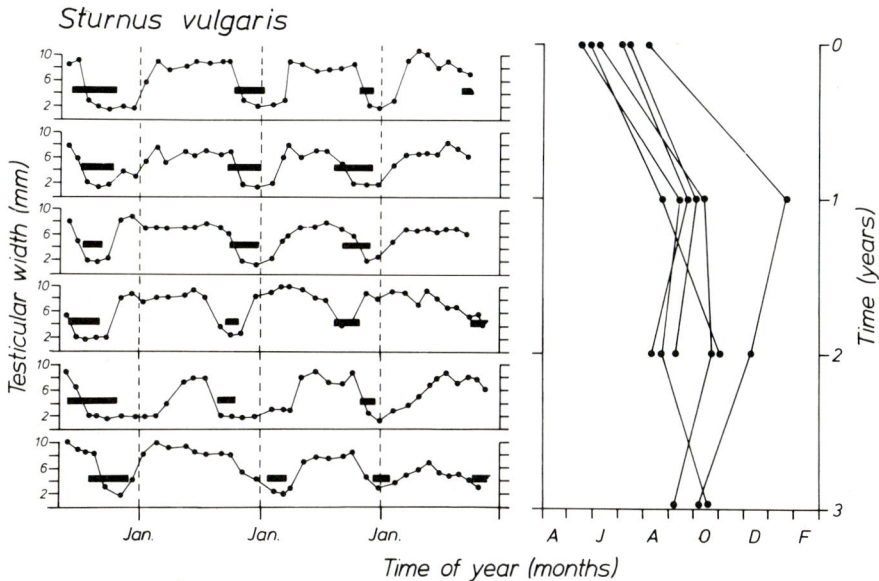

Fig. 2.12. Left Circannual rhythms in testicular width (*curves*) and molt (*black bars*) of six European starlings (*Sturnus vulgaris*) maintained for 43 months in a constant LD 11:11 (*upper three graphs*) or LD 12:12 (*lower three graphs*). *Right:* symbols connected by lines indicate the dates at which the individual birds began molting in the successive years of the experiment. (After Gwinner 1981b)

in either an LD 11:11 or in an LD 12:12. The free-running nature of these rhythms is again quite obvious (Gwinner 1981d).

Among birds, clear circannual rhythms have been found in 15 species from 5 families (Paridae, Sylviidae, Muscicapidae, Sturnidae, and Fringillidae, Table 2.1). In addition, there is suggestive evidence for seven other species (Table 2.2).

2.1.3 Lower Vertebrates

Relatively little is known so far about the existence of circannual rhythms in lower vertebrates. Still, it is clear that among reptiles a circannual rhythmicity plays a role in the control of activity in a lizard (*Sceloporus virgatus;* Stebbins 1963) and probably in a turtle (*Pseudemys scripta;* Hutton 1960; Table 2.2). The clearest demonstration of circannual rhythms in fish is that of behavioral thermoregulation in the white sucker (*Catostomus commersoni*) shown in Fig. 2.13 (Kavaliers 1982). Among certain salmonid fishes, growth rate and changes in body form and skin coloration related to the yearly smoltification cycle are most probably under circannual control (Brown 1945; Eriksson and Lundquist 1982, Table 2.2). The

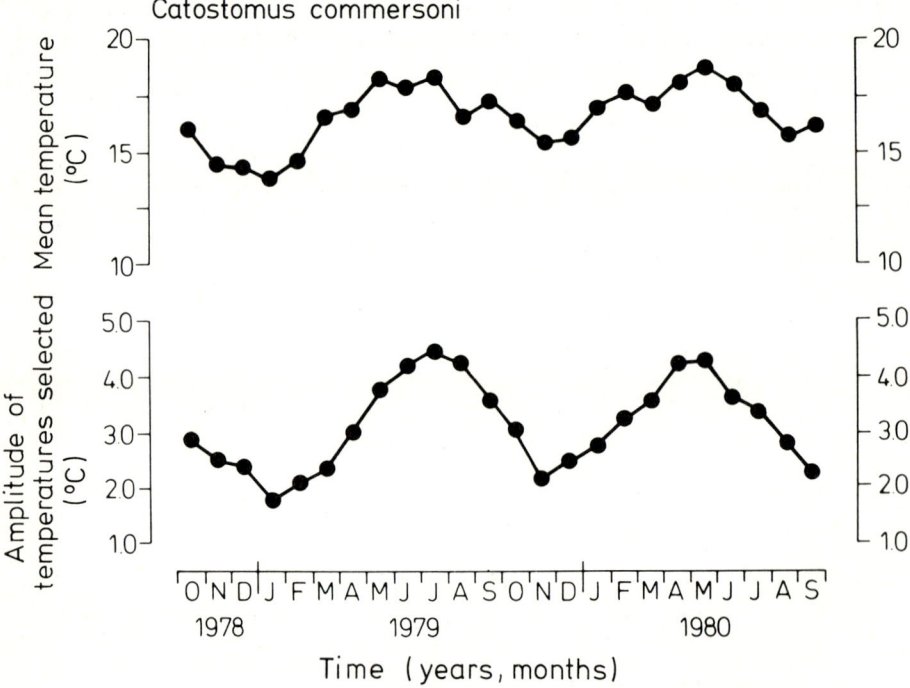

Fig. 2.13. Circannual rhythms of mean temperatures selected over 24 h and the amplitude of the diel pattern of temperatures selected by a white sucker (*Catostomus commersoni*) held in an LD 12:12 and a constant temperature of 15 °C. Each month the fish was exposed for 10 days to a temperature gradient of 4° to 20 °C where it was allowed to select its preferred temperature. (After Kavaliers 1982)

same holds true for the cycles of blood constituents and thyroid activity in the fish *Mystus vittatus* (Singh 1968) and for the ovarian cycle of the airsac catfish (*Heteropneustes fossilis;* Sundararaj and Vasal 1973; Sundararaj et al. 1982, Table 2.2).

2.1.4 Invertebrates

Early evidence of circannual rhythms among invertebrates came from results on carpet beetles (*Anthrenus verbasci;* Blake 1959a). These long-lived insects normally spend their first and second winter in larval diapause, after which they pupate, with emergence of adults during the following spring. When kept in DD and with constant temperature and humidity, the rhythms of diapause and pupation continue with a period between 10 and 11 months (Fig. 2.14). Although the period of these rhythms was hardly affected by temperatue, the proportion of beetles pupating during the first or second pupation peak depends on temperature. At 25° and 22.5 °C, under which development is rapid, all beetles pupate during the first pupation "gate". In contrast, at 15 °C pupation is delayed to the second, and, in one instance, to the fourth gate. Intermediate temperatures of 20° and 17.5 °C split the population, with some beetles pupating during the first, others during the second gate. This circannual "gating" rhythm bears a striking similarity to the circadian "gating" oscillations controlling pupal eclosion in *Drosophila pseudoobscura* and other insects (Pittendrigh 1981a, b). In these insects, the rate of devel-

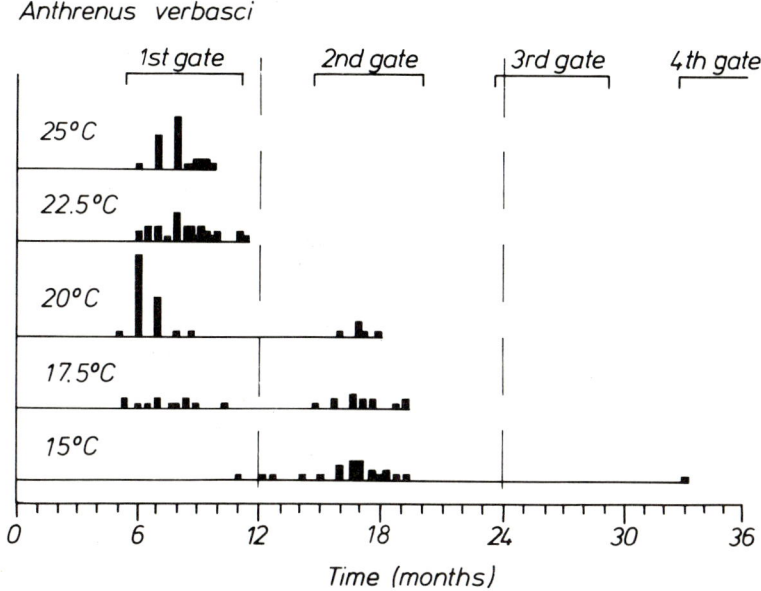

Fig. 2.14. Circannual rhythm of pupation in five populations of carpet beetles (*Anthrenus verbasci*) held in DD and at various constant temperatures. *Each black square* represents the pupation of an individual. (After Blake 1959a)

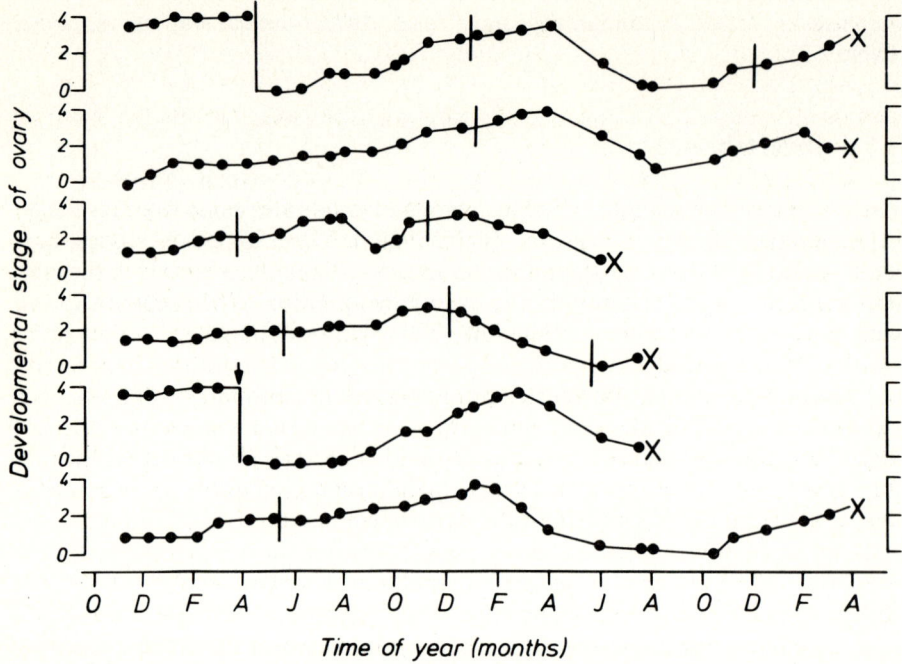

Fig. 2.15. Circannual rhythms of ovarian development (*curves*) and molt (*vertical lines on curves*) in six female cave crayfish (*Orconectes pellucidus*) held for up to 30 months in DD and at a constant temperature of 13 °C. ↓ Egg deposition; X animal died. (After Jegla and Poulson 1970)

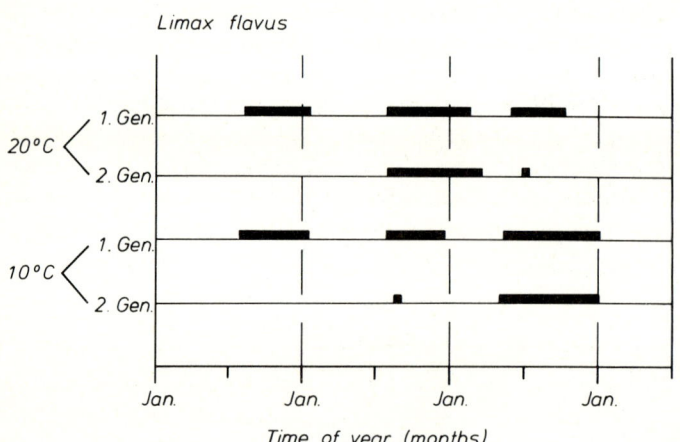

Fig. 2.16. Circannual rhythms of egg-laying in four groups of slugs (*Limax flavus*) held for 3 years in constant conditions of temperature and relative humidity (40°–65 °C) and under an LD 11:13. *Black bars* indicate periods of egg-laying. *1. Gen.*, *2. Gen.* First and second generation. (After Segal 1960)

opment depends heavily on temperature, but the circadian rhythm that determines eclosion is only marginally affected by temperature conditions.

Another arthropod for which a circannual rhythm has been demonstrated is the cave crayfish (*Orconectes pellucidus;* Jegla and Poulson 1970). Circannual rhythms in molt and the state of the reproductive organs were found in both males and females kept for 2.5 years in DD. The rhythms tended to be slightly shorter or longer than 12 months (Fig. 2.15).

Among molluscs, the slug *Limax flavus* has been shown to continue egg laying over three cycles in a constant LD 11:13 and constant temperature and humidity conditions (Fig. 2.16). The period was about 11 months at ambient temperatures of both 20° and 10°C (Segal 1960, Table 2.1). Similarly, snails of the species *Helix aspera* showed circannual changes in reproductive activity, food consumption, and locomotor activity when held for 14 months in LD 12:12 at 17.5°C (Bailey 1981, Table 2.2).

The most primitive animals for which circannual rhythms have as yet been documented are coelenterates, cnidarians of the species *Campanularia flexuosa*. When colonies of these animals were held for up to 39 months in DD and at 10°C, a rhythm in growth and development continued with a period slightly longer than 1 year (Brock 1975a–c, Table 2.1).

2.1.5 Plants

As mentioned in the introduction, although circannual rhythms have long been postulated by botanists, it still appears that "a rigorous demonstration of (circ)-annual rhythm in plants has probably never been achieved" (Sweeney 1969). Still, highly suggestive evidence is available from several studies. The most convincing data are on goose weed (*Lemna minor*) in which a rhythm in dry matter production continued for three cycles in cultures kept for 43 months in LL and constant temperature (Bornkamm 1966). The interval between the first and the second maximum tended to be slightly shorter than 1 year, whereas that between the second and the rather weakly expressed third maximum was clearly longer than 1 year (Fig. 2.17). These findings corroborate earlier suggestive findings in the same species (White 1936; Pirson and Göllner 1953). Another study in which more than one complete cycle was measured in constant experimental conditions is that of Kessler and Cygan (1963), who found circannual variations in nitrate reduction in cultures of the green alga, *Ankistrodesmus braunii* in LL at 23°C. The period of this rhythm was, however, indistinguishable from 1 year, and the seasonal time course of nitrate reduction was so similar during the two cycles that one has to question whether the rhythm was truly endogenous, or rather the result of uncontrolled environmental factors.

The capacity of certain seeds to germinate varies on an annual basis. In some of the early (Sperlich 1919; Okada 1930; Ruge and Liedtke 1951) and more recent studies (Lapeyronie 1968), however, these variations might have been due to changes in environmental factors that were not held constant throughout. The experiments of Bünning (1948, 1949b), Bünning and Müssle (1951), Bünning and Bauer (1952), in contrast, were carried out under rigidly controlled conditions

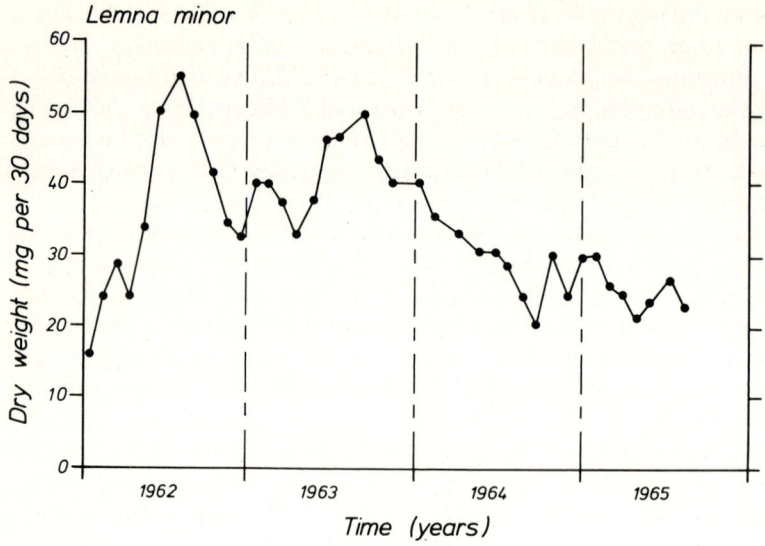

Fig. 2.17. Circannual rhythm in dry matter production of a culture of goose weed (*Lemna minor*) held for 43 months in LL and at 25 °C. (After Bornkamm 1966)

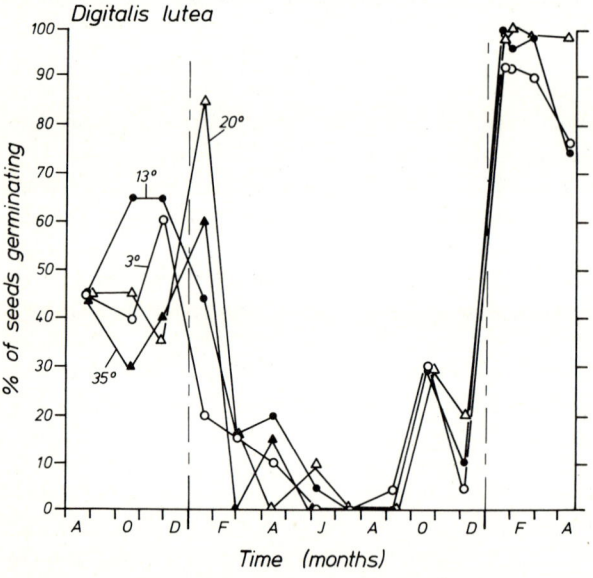

Fig. 2.18. Variations in the percentage of germinating seeds of *Digitalis lutea* stored in DD and at the various constant temperatures indicated. For the tests which were carried out at approximately 45-day intervals, samples of about 30 seeds were taken from the stock, and the percentage of seeds was determined which germinated within 2 weeks on moist filter paper at 23 °C, in constant light conditions. (After Bünning 1949b)

and therefore do suggest the participation of a circannual rhythmicity (Table 2.2). An example is shown in Fig. 2.18. Here the variations in the percentage of germinating seeds of *Digitalis lutea,* which had been stored in various temperatures, are plotted. A temperature-independent rhythm was indicated by these and several other results (Bünning 1949b; Bünning and Müssle 1951). Other experiments with *Gratiola officinalis* (Bünning and Müssle 1951) revealed that neither storage conditions (maintenance in air, O_2, CO_2 or N_2; Fig. 2.19a–d) nor an exposure of the seeds to high ambient temperatures for 1, 10 or 40 min at the beginning of the experiment (e–g) affected the period of the rhythm. In addition, slight or heavy mechanical damage applied to the seed husks before exposure to experimental conditions did not conspicuously alter the period of the rhythm (h, i). Puncturing the seeds before each test resulted in a dramatic increase in the percentage of seeds germinating, but a weak rhythmicity was still discernible (j). In another experiment Bünning (1949b) also demonstrated that humidity conditions during storage had only negligible effects on the period of the rhythms.

Bünning (1949b) obtained a remarkable result when he tested seeds of *Fragaria vesca* that had been harvested at four different times during the fruiting season (between early June and late July) and subsequently held at room temperature. Regardless of the harvesting time, the readiness to germinate increased si-

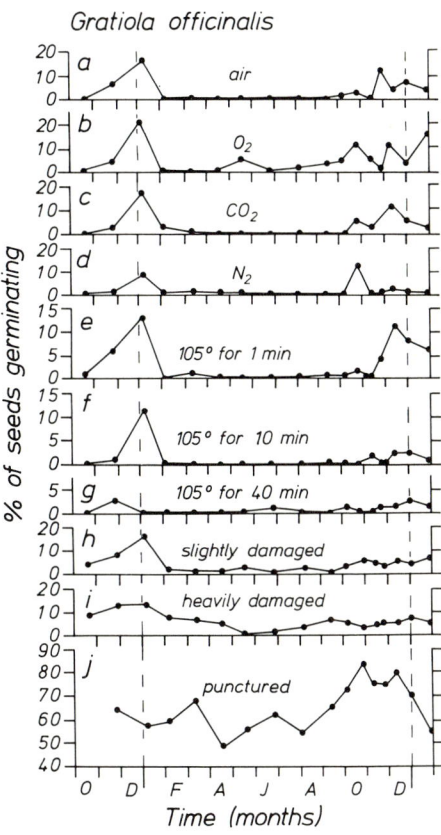

Fig. 2.19. Variations in the percentage of germinating seeds of *Gratiola officinalis* stored in DD at 20 °C. For the test, which was carried out at intervals of about 2 to 4 weeks, samples of about 100 seeds were taken from the stock and the percentage of seeds which germinated on moist filter paper at 25 °C and in DD was determined. *a–d* seeds stored in air, O_2, CO_2, and N_2. *e–g* seeds were exposed to 105 °C for 1, 10, or 40 min at the beginning of the experiment and subsequently stored in air. *h,i* seed husks were slightly or heavily damaged mechanically at the beginning of the experiment. *j* seeds were punctured by a needle just prior to each germination test. (After Bünning and Müssle 1951)

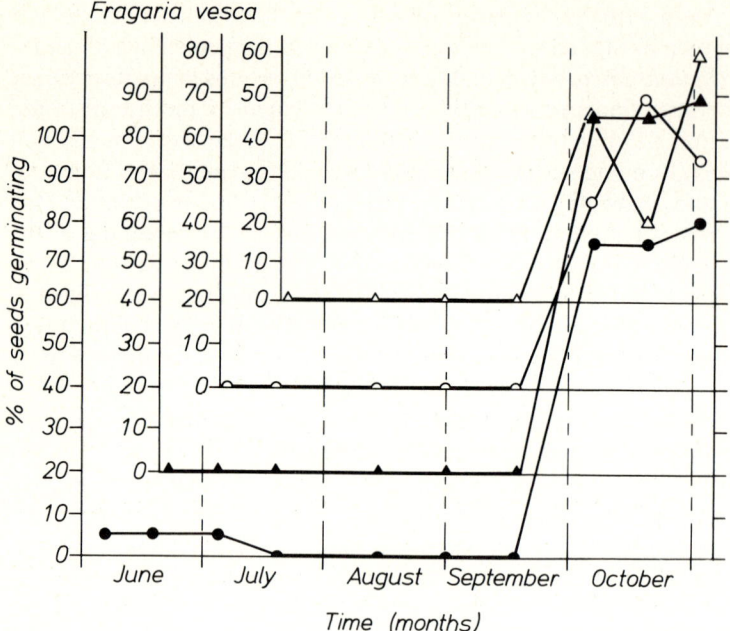

Fig. 2.20. Variations in the percentage of germinating seeds of *Fragaria vesca* harvested at four times during the fruiting season in early (●) and late (▲) June and in early (○) and late (△) July, and subsequently stored at room temperature. Germination tests at 2- to 3-week intervals were similar to those of the experiments shown in Figs. 2.18 and 2.19. The rapid increase in the capacity to germinate occurred simultaneously in all four samples. (After Bünning 1949b)

multaneously in early October in all samples tested (Fig. 2.20). This suggests that "the embryo takes over the endogenous annual rhythmicity from the mother plant so that it is in the same phase as the latter" (Bünning 1949b, p. 172, translated).

Taken together, these results on seed germination suggest that the phase of the rhythm in the seed is determined by the rhythm of the parent plant and that the period of the rhythm in the seed is homeostatically controlled, similar to the situation among circadian rhythms (see Chap. 3.3.1). Still, some doubts remain about whether a circannual rhythmicity was indeed involved. The problem is that more than one circannual period was apparently never measured under controlled conditions, and that the period of this first cycle was usually very close to 1 year. Moreover, in some experiments such as the one shown in Fig. 2.18, the various cultures held under different conditions became more synchronous as time progressed. The opposite would have been expected from free-running circannual rhythms. Hence, such data appear more consistent with the hypothesis of an exogenous induction of the observed rhythms by uncontrolled seasonal factors. Okada (1930) pointed out in line with this that the after-ripening of (*Euryale*) seeds "is a rather delicate process, slight modification of the environmental factors causing a remarkable influence upon it" (Okada 1930, p. 59).

The difficulties in rigorous demonstration of circannual rhythmicity in seeds is also demonstrated by the experiments of Kummerow (1963), who sealed fruits of the grass *Dactylis glomerata* in glass tubes and stored them in DD and at a temperature of either 5° or 25°C. Beginning in March 1960, germination tests were performed at about monthly intervals by placing caryopses on wet filter paper in an incubator at 24°C. During the 2-year experimental period the germination capacity went through two maxima, about 12 months apart. The rhythm was more clearly expressed in the samples stored at 5° than in those stored at 25°C. Unfortunately, another experiment carried out in 1958/59 with the same method and seeds harvested from the same piece of land produced rather different results. The seeds stored at 5°C showed no clear rhythm, whereas those stored at 25°C showed a rhythm, the wave form of which, however, was very different from that obtained in 1960/61. Moreover, in the 1958/59 experiment the water content of seeds stored at 25°C was low when germination capacity was high and vice versa, whereas in 1960/61 no relationship between these two parameters could be detected. These discrepancies are difficult to explain, but once again they point to the possibility that the observed variations in germination capacity may have been due to direct effects of external variations before or during the germination tests. This view is supported by the remark of Kummerow that one of his odd data points might have resulted from an elevation of the incubator temperature due to unusual environmental temperature conditions.

Results obtained on the seasonal variations in the formation of perennial organs in certain aquatic plants (e.g., Henssen 1954), in photomorphogenic reactivity of bean seedlings (Spruyt et al. 1983), and in circadian leaf-movement patterns of *Oxalis* (Müller-Haeckel 1975) are also not fully convincing because, as with most of the other studies, more than one cycle was not measured, and its period was always close to 12 months (Table 2.2). Despite these reservations, it appears that future studies with plants could be highly rewarding. A rigorous search for good plant models suited for the analysis of the mechanisms underlying circannual rhythms would seem to be a promising undertaking.

2.2 Atypical Cases

In the examples discussed so far the periods of the free-running rhythms were usually different from, but still relatively close to 12 months. There are, however, several reports of rhythms with periods that deviated to such an extent from 12 months that they can hardly be called circannual. Some of these instances are listed in Table 2.3. The cycles of testicular size described for Pekin ducks held for several years in LL or DD belong into this category (Fig. 2.21; Benoit et al. 1970). Of particular interest are organisms that show typical circannual cycles under some conditions, but atypically short ones in others. Common dormice, for instance, tended to show highly variable rhythms of body weight and hibernation with an average period of 162 days (range 28 to 425 days) if held in 5°C and in an LD 12:12 (Mrosovsky 1977), but developed rather consistent and typical circ-

Table 2.3. Summary of investigations providing evidence for "atypical" circannual rhythms with periods much shorter or longer than 12 months

	Species	Light conditions	Temperature °C	Maximal time in constant conditions (months)	Functions showing periodicity	Maximal number of complete cycles	Estimated period (months)[a]	Ref.
Fish	*Gasterosteus aculeatus*	LD 16:8	20	14	Reproductive activity	2	6.5 (♂) 7.5 (♀)	Baggerman (1957, 1980)
Birds	Pekin duck	LL	18	54	Testicular size	(5)	(4)	Benoit et al. (1955, 1956, 1959, 1970), Benoit (1970)
		DD		70		(7)	(8)	
	Erithacus rubecula	LD 8:16	20	30	Body weight	(2)	(15)	Merkel (1963)
		LD 18:6		16		(2)	(7)	
Mammals	*Spermophilus richardsoni*	LD 12:12	3	16	Body weight, food intake, molt, testicular state	2	6–8	Melnyk (1983)
	Spermophilus richardsoni	LD 12:12	0, 6, 12	33	Hibernation	4	7	Scott and Fisher (1970)
	Glis glis	LD 12:12	0	70	Hibernation	9	6.1	Scott and Fisher (1976)
	Glis glis	LD 12:12	5	19	Hibernation, body weight	8	5.4	Mrosovsky (1977)
			22				1.8	
	Glis glis	LD 12:12	23	15	Body weight	(10)	8.7	Melnyk (1979)
	Glis glis	LD 6:18	Constant between 13 and 23	36	Body weight, food and water intake, weight of liver, adrenal and salivary gland, testis weight, spermatogenic activity	18	1–7	Mrosovsky et al. (1980)
		LD 12:12						
		LD 18:6						
		LL						
	Glis glis	LD 12:12	6	12	Hibernation, body weight, plasma testosterone, plasma thyroxin	(4)	(3–6)	Jallageas and Assenmacher (1984)
	Eliomys quercinus	LD 12:12	22.5	24	Body weight	2	4–9[b]	Mrosovsky and Lang (1980)

[a] Numbers in parenthesis represent rough estimates.
[b] Two out of 14 animals, the others arrhythmic.

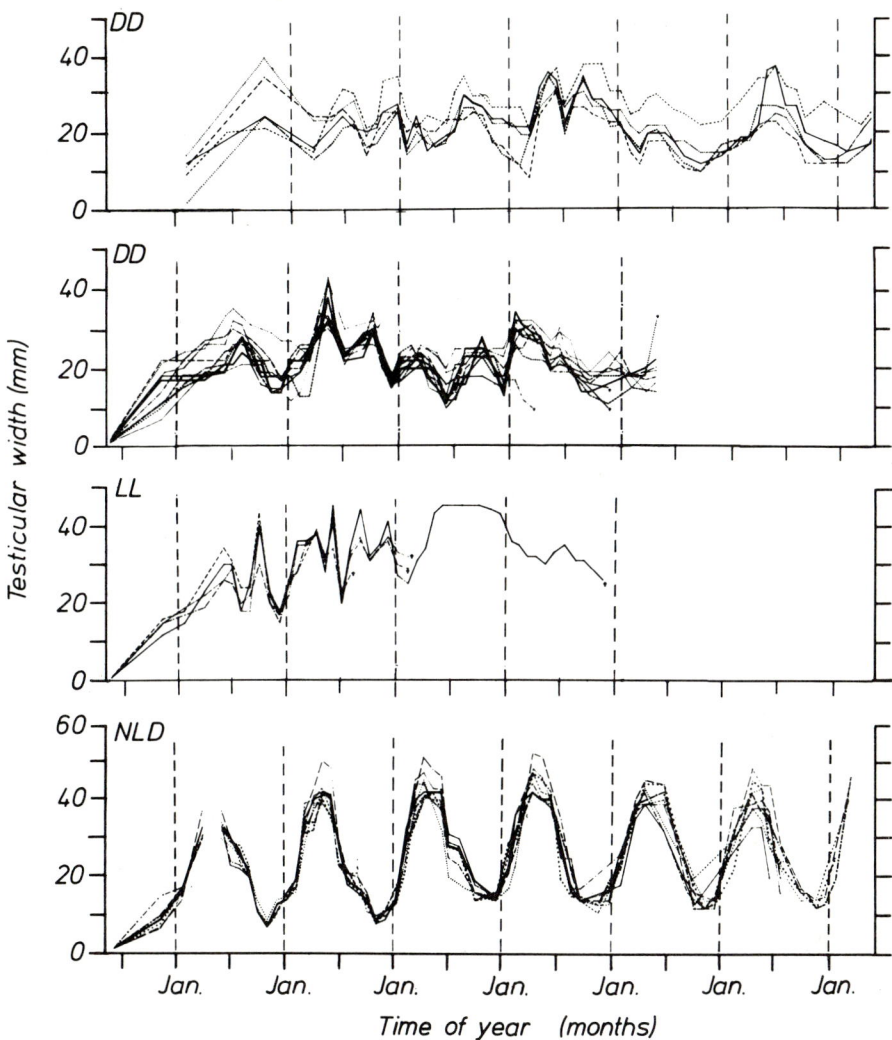

Fig. 2.21. Variations in the testicular width of Pekin ducks held for up to 7 years in DD, LL or under the natural photoperiodic variations of southern France. (After Benoit et al. 1970)

annual cycles with periods of 12 to 13 months if held in 12 °C and allowed to self-select between light and darkness (Fischer et al. 1975; Butschke 1977, Table 2.1). Occasionally rhythms have been described which under different environmental conditions have rather different and always "atypical" periods. The body-weight cycles of European robins (*Erithacus rubecula*) had periods of about 15 months if held in LD 8:16 but of about 7 months if held in LD 18:6 (Merkel 1963, Table 2.3). Finally, it has been observed relatively frequently that organisms go through one or two initial "typical" circannual cycles which are then replaced by irregular oscillations with much shorter periods (e.g., whitethroat, *Sylvia communis,* in LD 12:12, Gwinner 1983; willow warbler, *Phylloscopus trochilus* in LD 18:6, Gwinner 1971a). All of these "atypical" rhythms provide borderline cases of the subject treated in this monograph, which may, however, at some stage become of interest in connection with the physiological analysis of the mechanisms underlying circannual rhythmicities (for discussions see particularly Mrosovsky 1977, 1978, 1985; Mrosovsky et al. 1980).

Chapter 3

Properties of Free-Running Circannual Rhythms

3.1 Degree of Persistence and Range of Permissive Conditions

Circannual rhythms differ widely among species with regard to their tendency to persist or to dampen. The rhythms of hibernation in golden-mantled ground squirrels (*Spermophilus lateralis*), those of body weight and other parameters in an eastern chipmunk (*Tamias striatus*) and the cycles of molt in garden warblers (*Sylvia borin*) and blackcaps (*S. atricapilla*) have been found to continue for at least five, six, and nine cycles, respectively (Table 2.1). This means that the circannual clock can function essentially for the whole life of an organism (Pengelley and Asmundson 1969; Richter 1978; Berthold 1978). In other instances, circannual rhythms have shown more or less pronounced tendencies to dampen. This is true, for instance, of the hibernation rhythm in several species of ground squirrels and chipmunks (e.g., Pengelley and Kelly 1966; Heller and Poulson 1970), the body-weight rhythms of the dickcissel (*Spiza americana*: Zimmermann 1966) and the rhythms of molt and migratory disposition in the chiffchaff (*Phylloscopus collybita*: Gwinner 1971a) and whitethroat (*Sylvia communis*: Gwinner 1983). Particularly interesting from a physiological aspect are cases in which, under one particular constant condition, the annual rhythm of some functions persists in an organism whereas that of others fade away. Examples of this will be given in Chap. 5.

In many animals a circannual rhythmicity is expressed only within a very narrow range of environmental conditions. In the willow warbler (*Phylloscopus trochilus*), clear circannual rhythms in zugunruhe and molt were found in LD 12:12, but no such rhythm was observed in LD 18:6 (Gwinner 1971a). Similarly, blackcaps maintained in LD 10:14 and LD 12:12 showed rhythms of zugunruhe and molt for more than 3 years, whereas these rhythms disappeared after about 2 years in conspecifics maintained in LD 18:6 (Berthold et al. 1972a). The annual testicular cycle of the red-billed dioch (*Quelea quelea*) persisted for 2 years in LD 12:12, but apparently disappeared in LD 8:16 (Lofts 1962, 1964). Even more conspicuously, the European starling (*Sturnus vulgaris*) usually showed a well-defined circannual rhythm of testicular size when kept in LD 12:12, but no periodicity was observed in birds kept in photoperiods of 11 h or less or 13 h or more (Schwab 1971; Gwinner et al. 1985b). Sika deer (*Cervus nippon*), on the other hand, replaced their antlers with a circannual rhythmicity if exposed to constant light or to a constant 18-h or 6-h photoperiod, but these rhythms disappeared in a 12-h photoperiod (Goss 1969b). A narrow photoperiodic range of expression has also been found for the circannual rhythms of other mammalian species like the Turkish hamster (*Mesocricetus brandti:* Hall and Goldman 1982), but in hibernating mammals temperature conditions seem to be crucial as well. For instance, in the garden dormouse (*Eliomys quercinus*) held in LD 12:12 a circannual hibernation rhythm could be observed at 12 °C (Daan 1973) but not at 22 °C (Mrosovsky and Lang 1980).

Closely related species may differ conspicuously with regard to the conditions under which circannual rhythms are expressed. In the pied flycatcher (*Ficedula hypoleuca*), for instance, the circannual rhythms of gonadal size persisted in birds held in LD 12:12 but were arrested in an LD 12.8.:11.2. A sibling species, the collared flycatcher (*F. albicollis*), in contrast, remained cyclic under both conditions. This species' difference can be traced back to the fact that the latter species is capable of terminating photorefractoriness under both conditions, whereas for the pied flycatcher a photoperiod of 12.8 h is too long for termination of refractoriness (Gwinner 1986).

The decay of a rhythm is often the result of a steady reduction in its amplitude (e.g., that of body weight, gonadal size). In some instances such a reduction in amplitude is accompanied by a decrease in period length. This is true of circannual rhythms in testicular size of white-crowned sparrows (*Zonotrichia leucophrys*) in LD 12:12 in which no postnuptial molt and no photorefractoriness developed (Farner et al., unpublished; see Farner et al. 1983; Donham et al. 1983), and the rhythm of molt in whitethroats in LD 12:12 (Gwinner 1983).

There are animals that under some conditions show "typical" circannual rhythms, whereas under other conditions, these rhythms become "atypical" with much shorter and often rather variable periods. Common dormice (*Glis glis*) held in 12 °C under conditions in which they could select their preferred light intensity, exhibited a circannual rhythm of hibernation with a period of 12–13 months (Fischer et al. 1975), whereas conspecifics held at 5 °C in a constant LD 12:12 showed irregular hibernation cycles with an average period of about 5 months (Mrosovsky 1977). In common dormice, similar cycles with periods sometimes as short as 1 month have also been found in other studies and for other functions like body weight (Mrosovsky 1977; Melnyk 1979; Mrosovsky et al. 1980), plasma levels of testosterone and thyroxin (Jallageas and Assenmacher 1984), food and water intake; weight of the liver, adrenal and salivary gland and testes, and spermatogenetic condition (Mrosovsky et al. 1980; Joy et al. 1980). In meadow jumping mice (*Zapus hudsonius*) circannual rhythms in body weight and hibernation occurred under LD 12:12 at 5 °C, whereas body-weight rhythms with periods of about 3 months were observed in conspecifics held in the same photoregime at 20 °C (Muchlinski 1980). A replacement by or transition into high frequency oscillations of the "normal" rhythmicity under some conditions is also known from the circadian field (e.g., Gwinner 1966).

The observations mentioned above indicate a bewildering diversity of conditions under which circannual rhythms can be manifested. This is in sharp contrast to the rather uniform situation among circadian rhythms which, at least in most vertebrates, are usually suppressed at high intensities of illumination but continue at low light intensities (e.g., Aschoff 1979). The question of how this diversity in permissive conditions for circannual rhythms may be related to the ecology of the species studied and what it may tell us about the concrete mechanisms underlying circannual rhythms will be treated in subsequent chapters.

3.2 Range of Circannual Period Length and Transients

After transfer to constant conditions, circannual rhythms sometimes go through one or two transient cycles before they reach an apparent steady state (see for instance Figs. 2.8, 2.9, and 2.12). Unfortunately, in most studies only a few cycles have been measured, so that many of the conclusions drawn about circannual rhythms under constant conditions should be treated with caution, because it is usually not clear that they reflect steady-state properties of these rhythms.

Under constant conditions the periods (τ) of circannual rhythms are usually found to be shorter than 1 year, although longer periods also occur (Table 2.1). The overall range of individual periods measured during the second or a subsequent cycle in the typical cases summarized in Table 2.1, varies from about 6 to 16 months; the extreme values represent deviations from one year of about -50% and $+30\%$. If the atypical cases compiled in Table 2.3 are included, this range becomes even larger toward shorter periods.

Even within a single species, τ may show considerable *inter-individual* variations among animals maintained under identical conditions (Fig. 3.1, which is

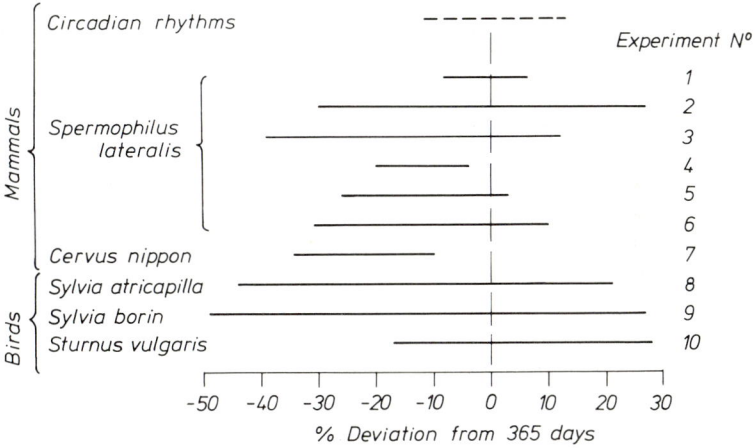

Fig. 3.1. Overall ranges of circannual period values (expressed as % deviation from 1 year) measured in animals of ten experiments in five species. *1* to *6* Golden-mantled ground squirrel (*Spermophilus lateralis*); *7* Sika deer (*Cervus nippon*); *8* Blackcap (*Sylvia atricapilla*): *9* Garden warbler (*Sylvia borin*); *10* European starling (*Sturnus vulgaris*). Only the second and the third cycle of free-running rhythms were considered. *Dashed horizontal line* shows the approximate overall range of most vertebrates circadian τs (Aschoff 1979), expressed as % deviation from one day. References: *1* Heller and Poulson (1970); *2* Pengelley et al. (1978); *3* Pengelley et al. (1976a); *4* Pengelley and Asmundson (1975); *5* Pengelley and Asmundson (1970); *6* Pengelley and Asmundson (1969); *7* Goss (1969b); *8, 9* Berthold et al. (1972a); *10* Gwinner (1981a)

based on experiments in which at least ten individuals were studied and only the values obtained during the second and/or third cycle in constant conditions were used). Compared with the overall range of free-running circadian period, the range of period of circannual rhythms is rather large, even within some of the individual experiments.

An *intra-individual* variability of circannual periods can be calculated from the few cases in which several successive circannual cycles were measured in individual animals. For the last five circannual cycles of hibernation in the eastern chipmunk studied by Richter (1978, Fig. 2.3), the standard deviation of the mean was 10 days, i.e., 3.1% of the mean of 330 days. This value is within the range of standard deviations measured for circadian rhythms (e.g., 1.3%–3.3%, and 1.9–4.2% of the mean circadian period of locomotor activity in the house mouse and the chaffinch, *Fringilla coelebs,* respectively; Aschoff 1981a). However, a considerably greater variability was found in the rhythm of molt of the two warblers shown in Fig. 2.9 (about 8.5 and 7% of the mean, respectively). Although other studies have measured too few cycles to calculate intra-individual variability, it appears from the inspection of many other data that the variability of period in the warblers is probably more representative of the majority of circannual rhythms than that of the eastern chipmunk.

In general it appears, then, that both inter- and intra-individual variability of τ is considerably larger in most circannual rhythms than in circadian rhythms, although there are cases in which circannual rhythms may show inter- and intra-individual variabilities similar to those known in circadian systems.

3.3 Dependence of Period on External Conditions

There are only a few studies in which the effects of external conditions on free-running circannual rhythms have been investigated. In most of these studies, only one or two cycles were reported, so the rhythms were perhaps not in steady state. The little evidence that is available from long-term experiments suggests that, once in steady state, the free-running circannual τ is often remarkably independent of external conditions.

3.3.1 Temperature

Among mammals there is some evidence that the periods of circannual cycles in body weight and hibernation of golden-mantled ground squirrels increases slightly with decreasing temperature. The most convincing data are those of Mrosovsky (1980b), who demonstrated that animals held for 34 months in an LD 12:12 at 9.5 °C had a considerably longer period, almost 2 months, in the cycle of body weight than individuals held at 21 °C (Fig. 3.2). These results corroborate earlier, less conclusive findings from the same species (Pengelley and Fisher 1963) in which squirrels were held in groups of four at 26 °, 22 °, and 0 °C. The first pe-

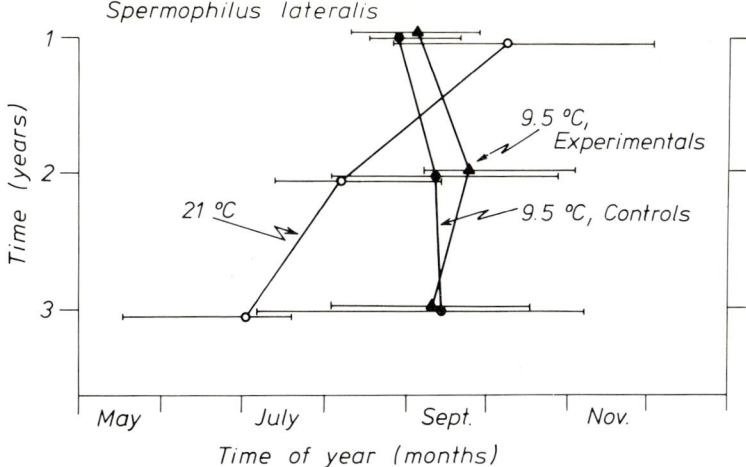

Fig. 3.2. Medians with ranges of maximum body weights in three groups of golden-mantled ground squirrels (*Spermophilus lateralis*) held for nearly 3 years in a constant LD 12:12 at constant temperatures of 9.5° or 21 °C. The animals of the 9.5 °C experimental group were deprived of food for about 6 months between November of the first and May/June of the second year, which resulted in an extension of the hibernation period by about 77 days, compared with the 9.5 °C controls. However, the subsequent occurrence of the maximal body weight was only marginally affected by this treatment. (After Mrosovsky 1980b)

riod of their circannual body-weight rhythms was 251, 304, and 342 days, respectively. Moreover, Pengelley and Kelly (1966) and Pengelley and Asmundson (1969) found that the first period of the circannual hibernation rhythm tended to be slightly longer in animals held under 3 °C than in animals held under 12 °C [$\tau_{3°} = 361$ days, $\tau_{12°} = 333$ days, Pengelley and Kelly 1966; $\tau_{3°} = 366$ days, $\tau_{12°} = 337$ days, Pengelley and Asmundson 1969]. In the two latter studies the differences were at the borderline of statistical significance. The difference disappeared, however, during the second and third cycle (Pengelley and Asmundson 1969). Slightly, but not significantly longer (344 days) circannual cycles in body weight were also found in golden-mantled ground squirrels held at 5 °C as compared with animals held under 34 °C (331 days; Pengelley et al. 1978). In contrast to these findings, no differences in the circannual period of hibernation were found between golden-mantled ground squirrels held by Scott and Fisher (1970) in 0° and 18 °C, respectively. In addition to the results from ground squirrels, there is suggestive evidence for a lengthening effect of lower temperatures in the eastern chipmunk (Scott and Fisher 1972).

It should be emphasized that, although these data on hibernators indicate that lowering ambient temperature lengthens the circannual period, the effect is relatively small. If held in low temperatures, these hibernators spend many months each year in deep torpor with body temperatures close to environmental temperature. Consequently, much larger effects would have to be expected if the processes underlying circannual rhythms showed a temperature dependence of the magnitude known for many other physiological processes, including the period of the high-frequency body weight cycles of common dormice (Mrosovsky 1977;

Mrosovsky et al. 1980). Hence these data suggest that the circannual period is to some extent temperature compensated.

More conspicuous examples of temperature-compensated circannual rhythms are known from lower organisms, of which the annual eclosion rhythm of the carpet beetle (*Anthrenus verbasci;* Fig. 2.14) is an excellent example. The circannual rhythm of oviposition of the slug (*Limax flavus*) also seems to be largely independent of temperature (Segal 1960, Fig. 2.16). Finally, several of the studies with plants, particularly with seeds, suggest temperature compensated circannual mechanisms (Fig. 2.18, Chap. 2).

3.3.2 Photoperiod

Table 3.1 summarizes results of experiments in which the influence of photoperiod on circannual τ have been investigated. In most of these studies no significant effects could be detected. Among mammals only the experiments of Pengelley et al. (1976a) with golden-mantled ground squirrels (No. 9 in Table 3.1) indicate that the periods of the second and third circannual hibernation cycle were significantly shorter in continuous light than in LD 12:12. In other studies with ground squirrels, however, no effects of photoperiod on τ could be detected (Nos. 8 and 10), nor did bilateral removal of the eyes alter circannual period length (Pengelley and Asmundson 1970; Pengelley et al. 1976b; see Fig. 2.2). A (statistically nonsignificant) tendency for longer τ in LL as compared with τ measured in both LD 8:16 or 16:8 is suggested for the second circannual cycle of antler replacement in the Sika deer (No. 11). Ewes (*Ovis aries*) also tended to have a longer circannual τ of breeding activity in LL than in LD 8:16, LD 12:12 or LD 18:6, although none of the differences was significant (No. 13). Among birds, it was found that the second and third circannual period of molt, zugunruhe and body weight of garden warblers is significantly shorter in birds held in LD 12:12 than in birds held in 16:8. During the first cycle the circannual τ of warblers held in LD 16:8 was also significantly longer than in birds held in LD 10:14 (No. 2). In addition, some significant differences were found between the circannual periods of testicular size of garden warblers and of testicular size and molt of starlings exposed to light-dark cycles with periods differing from 24 (Nos. 4 and 5).

The meaning of these few photoperiodic effects is not clear. The pronounced differences in most of the experiments listed in Table 3.1 between the periods measured during the first and subsequent cycles suggest that the rhythms may still have been in a transient state, even during the second and third cycle. Truly long-term studies on rhythms in steady state are necessary to resolve this problem. Consequently the question of whether there are any rules in the dependence of circannual period on length of photoperiod, comparable to the effects of light intensity on circadian period length (Aschoff 1979) remains unanswered.

Table 3.1. Circannual periods under different lighting conditions

	No.	Species	Photoperiod	Period (days)[a]		Function	Ref.
				1st cycle	2nd cycle or 2nd and 3rd cycle		
Fish	1	*Heteropneustes fossils*	DD LL	(400) (400)		Ovarian weight	Sundararaj and Vasal (1973)
Birds	2	*Sylvia borin*	LD 10:14 LD 12:12 LD 16: 8	359 ± 30.3 (8) ←[c] 357 ± 24.8 (7) -- 391 ± 34.6 (6) --→	325 ± 46.3 (8) 309 ± 57.5 (7) 350 ± 57.3 (6)	Molt, zugunruhe, body weight	Berthold et al. (1972a)
	3	*Sylvia atricapilla*	LD 10:14 LD 12:12 LD 16: 8	351 ± 32.5 (6)[d] 355 ± 37.6 (7) 345 ± 31.6 (4)	308 ± 57.4 (6) 314 ± 47.8 (7) —	Molt, zugunruhe	
	4	*Sylvia borin*	LD 11:11 LD 12:12 LD 13:13	297 ± 11 (6) ←→ 410 ± 19 (5) ←×→ 362 ± 23 (5)		Testicular width	Gwinner (1981a)
	5	*Sturnus vulgaris*	LD 11:11 LD 12:12 LD 11:11 LD 12:12	468 ± 34 (9) ←→ 402 ± 60 (5) ←-- 462 ± 36 (9) --→ 443 ± 51 (7)	395 ± 112 (6) 328 (2) 395 ± 130 (8) 465 ± 134 (6)	Testicular width Molt	
	6	*Zonotrichia leucophrys*	LD 8:16 LD 20: 4	(365) (6) (365) (6)		Body weight	King (1968)
	7	*Lonchura punctulata*	LD 12:12 LL	(300) (5) (300) (5)	(330) (330)	Testicular size	Chandola et al. (1982)
Mammals	8	*Spermophilus lateralis*	LD 12:12 LD 20: 4	416 ± 67.3 (5) 356 ± 15.6 (5)	305 ± 7.8 (3) 307 ± 28.7 (4)	Hibernation	Pengelley and Asmundson (1970)
	9	*Spermophilus lateralis*	LD 12:12 LL (500 1×) LL (20 1×)	334 ± 26.3 (9) ←↑ 345 ± 28.4 (9) --↓ 354 ± 33.1 (10) --→	327 ± 27.7 (9) 305 ± 18.2 (8) 298 ± 35.7 (9)	Hibernation	Pengelley et al. 1976a
	10	*Amospermophilus leucurus*	LD 12:12 LL	420 ± 71 (6) 358 ± 28 (4)		Testicular size	Kenagy (1981b)

Table 3.1 (continued)

No.	Species	Photoperiod	Period (days)[a]		Function	Ref.
			1st cycle	2nd cycle or 2nd and 3rd cycle		
11	*Cervus nippon*	LD 8:16 LD 16: 8 LL	370 (1) 381 (3) 318 (4)	309 (1) 297 (2) 373 (3)	} Antler replacement	Goss (1969b)
12	*Ovis aries*	LD 8:16 LD 16: 8		315 ± 158 (6)[b] 292 ± 71 (6)	} Testicular size	Howles et al. (1982)
13	*Ovis aries*	LD 8:16 LD 12:12 LD 18: 6 LL	169 (10) 206 (10) 229 (10) 243 (10)	375 (10) 318 (10) 371 (10) 418 (10)	} Breeding activity	Ducker et al. (1973)

↕ $p < 0.01$ ↑↓ $p < 0.05$.

[a] Means ± S.D.; number in parenthesis: n.
[b] Combined first and second cycle according to time series analysis.
[c] Calculated from period values of the circannual rhythms of molt, zugunruhe and body weight.
[d] Calculated from period values of the circannual rhythms of molt and zugunruhe.

3.3.3 Social Factors

The only indication that social cues might affect circannual τ comes from a study with European starlings. If held in LD 12:12, the first circannual cycle of testicular size was significantly shorter in males held in individual cages than in males held together with other males or females. This difference disappeared, however, during later cycles (Gwinner 1981a). In the same birds it was also shown that the pattern of testicular growth and regression was affected by the presence of females (Gwinner 1975a).

3.4 Innateness of Circannual Rhythms

Although evidence is scare, there are results indicating that circannual rhythms may develop in animals that have never been exposed to annual environmental variations. Most convincing are demonstrations of circannual rhythms in animals already born in a seasonally constant environment. Golden-mantled ground squirrels, born and subsequently held in LD 12:12 and constant temperature conditions, exhibit circannual rhythms in hibernation indistinguishable from those of conspecifics caught in the wild (Heller and Poulson 1970; Pengelley and Asmundson 1970). Individuals of the same species born and then held in LD 14:10 under constant temperatures showed circannual rhythms in body weight and the plasma levels of testosterone and LH, respectively (Licht et al. 1982; Zucker and Boshes 1982). Similar results were obtained for the annual cycles of reproductive activity in ewes raised from birth under various constant photoperiods (Ducker et al. 1973). A circannual rhythm in egg deposition of slugs (*Limax flavus*) was observed in animals whose parents had already been living in constant conditions for about a year (Segal 1960). In addition, there are many reports on circannual rhythms in animals that had been transferred to constant conditions at an age of only a few months or weeks (e.g., several species of warblers: Gwinner 1971a, 1983; Berthold et al. 1972a; Berthold 1974a, b; flycatchers: Gwinner and Schwabl-Benzinger 1982; stonechats: Gwinner and Dittami 1985; sheep: Howles et al. 1982). Taken together, these data suggest that circannual rhythms do not become imprinted on animals during ontogeny by seasonally varying external factors, but represent innate characteristics of the organisms that develop independently of relevant environmental information.

3.5 Comparison with Circadian Rhythms

Richter (1978) concluded from his results with eastern chipmunks that "the yearly clock has all the characteristics of the 24-h clock"; Bünning (1956) states even more explicitly that "the phenomena of endogenous annual rhythmicity in plants

Table 3.2. Comparison between general properties of circadian and circannual rhythms

	Circadian	Circannual
Light conditions under which rhythms persist	Low light intensities	Very variable among species
Overall range of free-running periods	Small	Large
Inter-individual variability	Small	Large
Intra-individual variability	Small	Large
Effect of temperature on τ	Small – according to rules	Small – so far no rules
Effect of light intensity or photoperiod on τ	Small – according to rules	Small – so far no rules
Innate	Yes	Yes

and animals, as well as their regulation by external factors, are so similar that one might suspect a common physiological origin" (translated). On the basis of the information given in the foregoing paragraphs it appears, however, that such general statements are not justified although there are some similarities between these two rhythmic phenomena (Table 3.2). As in most circadian rhythms, the period of at least a few circannual rhythms seems to be temperature-compensated, but it is not clear as yet whether there are rules about the dependence of circannual τ on temperature comparable to those known about temperature effects on circadian τ (Aschoff 1979). Lighting conditions appear to affect circannual τ in a few species, but, once again, nothing can be said about whether any generalizations are possible as in the circadian field (Aschoff 1979). There is evidence for a few species that circannual rhythms, like circadian rhythms, are innate mechanisms rather than the result of imprinting or learning. With regard to other features, however, circannual rhythms differ widely from circadian rhythms (Table 3.2). The overall relative range of periods shown in constant conditions is usually much larger and intra- and inter-individual variability is higher in most circannual systems investigated than in circadian rhythms. Finally, and most conspicuously, the conditions under which circannual rhythms express themselves are extremely variable. Some species show rhythms under a very wide range of conditions, whereas others show them only under limited and very specific circumstances. These and other properties, which will be discussed later (Chap. 5), suggest that circannual rhythms are probably rather heterogenous in origin, and, consequently, that the mechanisms underlying them may be equally variable among species.

Chapter 4
Synchronization of Circannual Rhythms

4.1 Zeitgebers

Under seasonally constant conditions circannual rhythms usually free-run with periods somewhat different from 12 months as discussed in the previous chapters. Under natural conditions, on the other hand, their period normally matches that of the natural year. This implicates the existence of zeitgebers, i.e., seasonally varying environmental factors that are capable of synchronizing (entraining) circannual rhythms with the yearly cycle of the seasons.

The action of such circannual zeitgebers is illustrated by the results of a displacement experiment carried out on the woodchuck (*Marmota monax*), a species with an annual cycle in body weight that depends on a circannual rhythmicity (Table 2.1). In the experiment summarized in Fig. 4.1, woodchucks were first kept in the United States at about 40° northern latitude, but were then displaced to Sydney, Australia, about 34° south. This displacement corresponds to a phase shift of the seasonal environmental cycles by 180°. It can be seen that the rhythm in body weight gradually followed this phase shift, although it took more than

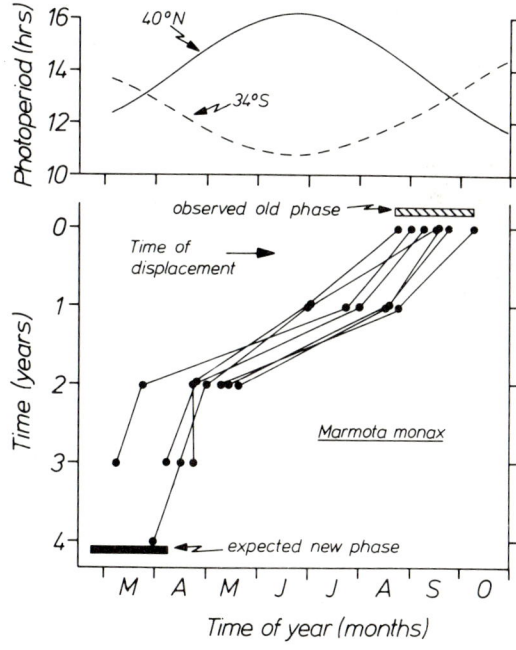

Fig. 4.1. Resynchronization of the circannual rhythms of body weight in seven woodchucks (*Marmota monax*) after displacement from the northern hemisphere (Pennsylvania, USA, about 40 °N) to the southern hemisphere (Sydney, Australia, about 34 °S). At both locations the animals were exposed to natural conditions of photoperiod and temperature. *Upper diagram* photoperiodic variations to which the animals were exposed before (*solid curve*) and after (*dashed curve*) displacement. *Lower diagram:* symbols connected by lines indicate the dates at which the animals attained maximal body weights before (year 0) and after (years 1–4) displacement. (After Davis and Finnie 1975)

one cycle until the expected phase relationship to local calendar time became re-established.

Although this type of experiment demonstrates the existence of circannual zeitgebers, it gives no insight into the concrete nature of the factors involved. To investigate this question, organisms must be held under conditions in which only one environmental factor is manipulated with all the others being held constant. Basically, four types of experiments can be carried out to test whether an environmental variable is a zeitgeber of a circannual rhythmicity (Aschoff 1960; Hoffmann 1969; see also Appendix).

1. *Exposure of a free-running rhythm with period τ_n to the environmental cycle, with period T.* If the latter is effective as a zeitgeber, the endogenous rhythm should assume its period so that $\tau = T$. After removal of the environmental cycle the endogenous rhythm should free-run again with its natural period τ_n.

2. *Varying the period of the environmental cycle.* In this case the period τ of the biological rhythm should follow changes in the period of the environmental cycle within certain limits.

3. *Phase shifting the environmental cycle.* If it is a zeitgeber, the biological rhythm should follow that phase shift within some cycles.

4. *Exposure of animals kept under constant conditions to pulsatile or stepwise changes of an environmental variable.* If it is a zeitgeber, the pulse or step should induce a phase shift, the size and direction of which should depend on the circannual phase exposed to the stimulus. It must be emphasized, however, that the existence of such a phase response curve is only a necessary but not a sufficient condition for the function of this variable as a zeitgeber, because entrainment will only be possible if the phase-response curve has certain properties (Pittendrigh 1981a).

In searching for circannual zeitgebers, mainly procedure (2) has been employed so far, although a few phase-shift experiments [procedure (3)] have been carried out as well; furthermore some evidence comes from data suggesting the existence of phase-response curves [procedure (4)]. From the available results, it appears that the annual cycle of photoperiod is the most powerful zeitgeber for circannual cycles, at least in most vertebrate species, but seasonal temperature cycles also seem to play a role both among vertebrates and evertebrates. In addition, there is one report suggesting that social stimuli may have zeitgeber qualities.

4.1.1 Photoperiod

Figure 4.2 provides a first example demonstrating that the annual cycle of photoperiod is capable of synchronizing the circannual rhythms in gonadal size and molt in an avian species, the European starling (*Sturnus vulgaris*). Two groups of starlings were exposed to sinusoidal changes in photoperiod mimicking in their general shape and amplitude those occurring at 40° latitude. They differed from each other only with regard to their period length. In one group the period of the photoperiodic cycle was T = 12 months, in the other T = 6 months. It can be seen in the upper diagram that the birds exposed to a 12-month cycle went through

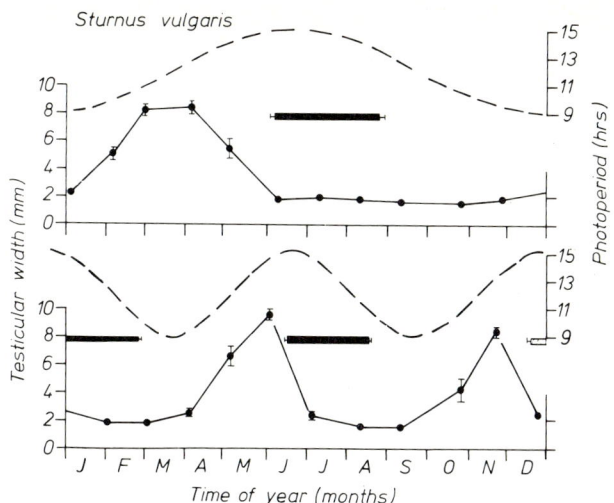

Fig. 4.2. Variations in testicular width (*solid curves*) and occurrence of molt (*black bars*) in two groups of European starlings (*Sturnus vulgaris*) exposed to the variations in photoperiod indicated by the *dashed curves*. The period of the photoperiodic cycle in the *upper diagram* was 12 months and in the lower 6 months. *Vertical lines at the symbols* and *horizontal lines at the bars* indicate standard deviations. (After Gwinner 1977a)

testicular growth and regression as photoperiod increased, followed by a complete molt. The lower diagram indicates that a similar relationship between the environmental photoperiodic cycle and the rhythms in testicular size and molt was also established in the birds held under a 6-month photoperiodic cycle. These birds went through the same sequence of events but this time within 6 months. In effect the circannual rhythms of these birds remained synchronized to a photoperiodic year reduced to half its normal duration.

Even photoperiodic cycles much shorter than 6 months are capable of entraining the starlings circannual rhythms (Fig. 4.3, left). Here starlings were exposed to photoperiodic cycles with periods of $T=12$, $T=8$, $T=6$, $T=4$, $T=3$, $T=2.4$, $T=2.0$, $T=1.7$, and $T=1.5$ months. It can be seen that photoperiodic cycles as short as 2.4 months still synchronized both the testicular and molt rhythms. Only in cycles even shorter than that signs of irregularities became apparent (see p. 61 ff.).

In a similar set of experiments Goss (1969a) exposed sika deer (*Cervus nippon*) to sinusoidal photoperiodic cycles with periods ranging from $T=2.0$ to $T=24.0$ months. The results (Fig. 4.4) indicate that photoperiodic zeitgebers with periods between $T=12$ and $T=4$ months synchronized the circannual rhythm of antler growth and shedding in all individuals. Under longer or shorter periods, one or several zeitgeber cycles were skipped by at least some animals, suggesting again that the circannual rhythmicity follows changes in the zeitgeber period only within certain limits (see below).

Table 4.1 summarizes results of experimental studies in which the effects of different photoperiodic cycles on circannual systems have been investigated. It shows that not only sinusoidal changes, but also rectangular and symmetrical as well as asymmetrical saw-tooth cycles, can be effective photoperiodic zeitgebers for a variety of different circannual functions in a variety of different species.

Of particular interest is the demonstration that the circannual rhythm of antler growth in sika deer could even be synchronized with stepwise increases and

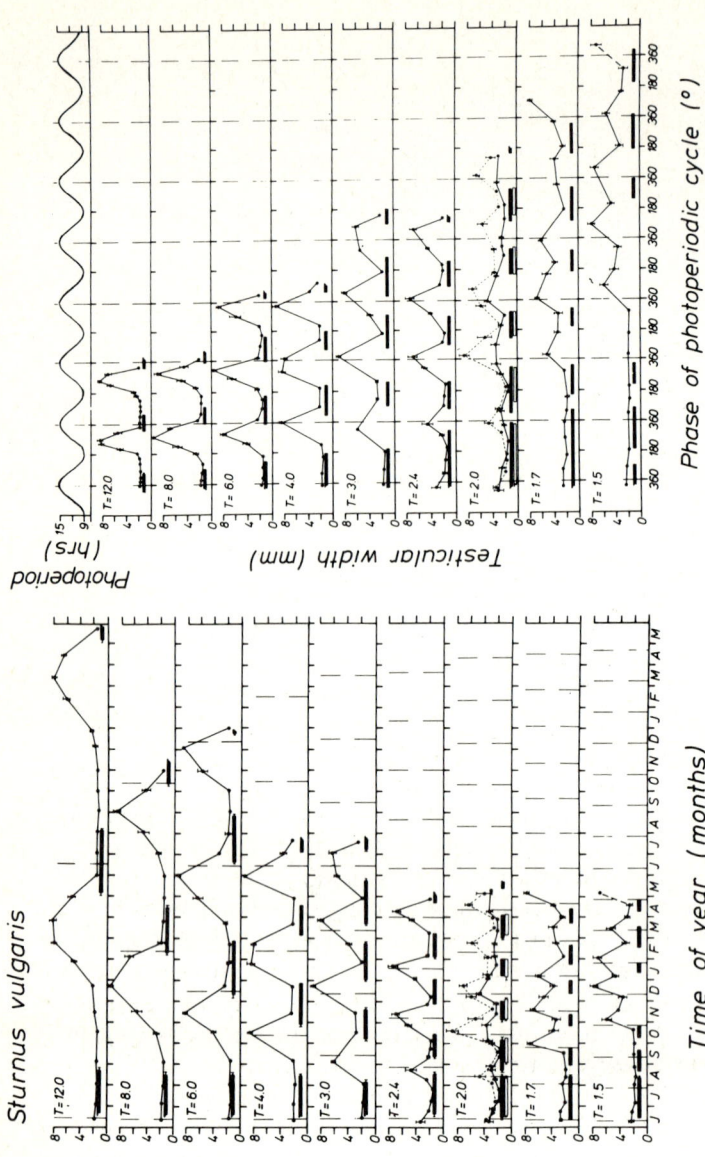

Fig. 4.3. Variations in testicular width (*curves*) and occurrence of molt (*bars*) in ten groups of European starlings (*Sturnus vulgaris*) exposed to sinusoidal changes in photoperiod. The amplitude and the general shape of these cycles were the same in all groups (and identical with those occurring at latitude 40°) but their duration varied from T = 12 months (*uppermost panel*) to T = 1.5 months (*lowermost panel*). On the *left* the data are plotted relative to the time of year and on the *right*, relative to the phase of the photoperiodic cycle, the total duration of which was normalized to 360° for all groups. *Dashed vertical lines* represent times of longest photoperiods ("photoperiodic mid-summer"). Under T = 2.0 two replicate experiments were performed; the results of one of them are presented as *dashed curves* and *open bars*. *Vertical lines at the symbols and horizontal lines at the bars* indicate standard deviations. (After Gwinner 1981b)

52

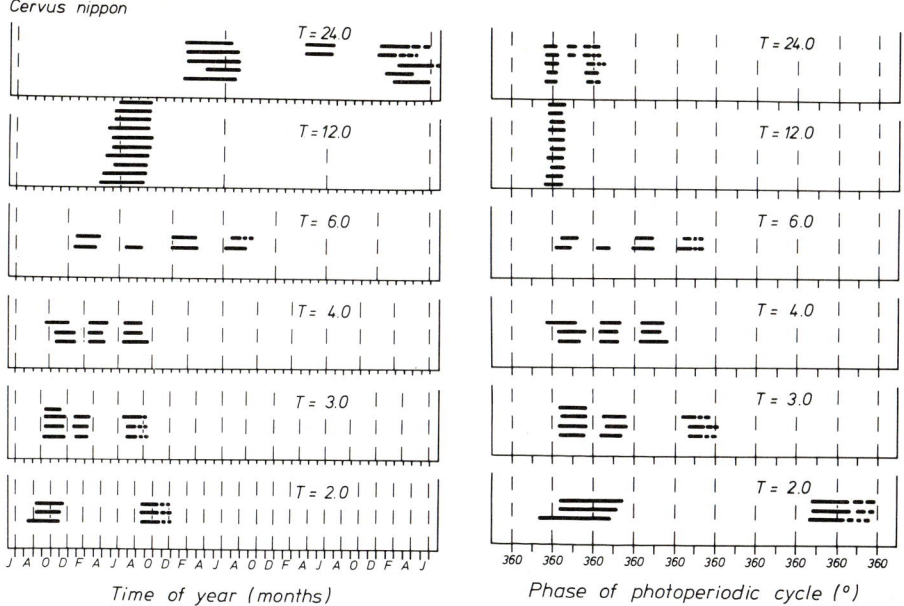

Fig. 4.4. Phases of antler growth (*bars*) in individual sika deer (*Cervus nippon*) exposed to sinusoidal changes in photoperiod. The amplitude and general shape of these cycles were the same in all groups (and identical with those occurring at latitude 40 °N) but their duration varied from T = 24 months to T = 2.0 months. On the *left* the data are plotted relative to the time of year and on the *right* relative to the phase of the photoperiodic cycle the total duration of which was normalized for all groups to 360°. *Dashed vertical lines* represent times of longest photoperiods. (Goss 1969a, after Gwinner 1981b)

decreases in photoperiod. In these experiments one group of deer was initially exposed to a 20-h photoperiod which was then reduced by 2 h every 4 months until a 4-h photoperiod was reached 3 years later. The reciprocal group was initially exposed to a 4-h photoperiod which was increased by 2 h every 4 months, until a 20-h photoperiod was attained. Figure 4.5 shows that during the first part of the experiment animals of both groups responded to every other change in photoperiod with antler replacement, irrespective of whether photoperiod increased or decreased. This result is perhaps most easily explained by the hypothesis that the relevant stimulus of a photoperiodic zeitgeber is "change in photoperiod" regardless of the direction of the change, and that the circannual event locks on to every second change. No other similar case has yet been described. Still, such a mechanism could explain entrainment to the natural photoperiodic cycle, which also changes twice a year. Deer typically replace their antlers at every second photoperiodic change. The situation is, however, not that simple, as indicated by the behavior of the deer during the second part of the experiment. Here, the regular pattern observed initially became highly distorted for some undetermined reason.

In the studies discussed so far the effectiveness of photoperiod as a zeitgeber has been demonstrated by changing the period of the photoperiodic cycle accord-

Table 4.1. Synchronization of circannual rhythms with photoperiodic cycles of different shapes and periods

	Species	Circannual function measured	Shape of photoperiodic cycle	Range (h)	T (months)	Entrainment Full	Entrainment Cycles skipped	By frequency division	
Birds	*Sylvia borin*	Zugunruhe, body weight, molt	~	9.0–18.0[a]	6.0	x[a]		x[a]	Berthold (1979a)
	Sylvia borin	Zugunruhe, body weight, testicular size,	~~	9.0–15.0	6.0	x			Gwinner (unpublished)
					12.0	x			
	Sturnus vulgaris	Testicular size, molt	~	9.0–15.0	12.0	x			Gwinner (1977a, 1981b)
					8.0	x			
					6.0	x			
					4.0	x			
					3.0	x			
					2.4	x			
					2.0	x			
					1.7		x		
					1.5		x		
	Sturnus vulgaris	Testicular size, molt	~		12.0	x			Gänshirt und Gwinner (1979)
Mammals	*Microcebus murinus*	Body weight, testicular size	~~~	8.1–16.1	12.0[b]	x			Petter-Rousseaux (1972)
				11.0–13.2	12.0	x			
				11.0–13.2	6.0	x			
	Microcebus murinus	Testicular size, estrous	zigzag	6.0–8.0	6.0	x			Petter-Rousseaux (1975)
				10.3–14.0	3.0	x			
	Cricetus cricetus	Body weight	~~	8.0–16.0	12.0[b]	x			Canguilhem et al. (1973)
				8.0–16.0	6.0	x			
	Cervus nippon	Antler shedding	⊓	9.0–15.0	24.0	x[c]		x[c]	Goss (1969a)
					12.0	x			
					6.0	x			
					4.0	x			
					3.0		x		
					2.0			x	
				9.0–15.0	4.0	x			

Species	Parameter		LD cycle (h)	Period		Reference
Cervus nippon	Antler shedding		[c]	8.0[d]	x	Goss (1976)
Ovis aries	Estrous		8.0–16.0	12.0	x	Rougeot (1961)
				6.0	x	
Ovis aries	Estrous		8.0–16.0	6.0	x	Rougeot (1962)
				4.0	x	
				3.0	(No rhythm)	
			10.0–14.0	6.0	x	
			11.0–13.0	6.0	x	
Ovis aries	Estrous		8.0–16.0	12.0	x	Mauléon et Rougeot (1962)
Ovis aries	Vaginal dilatability		8.0–22.0	12.0	x	Wodzicka-Tomaszewska et al. (1967)
				8.0	x	
Ovis aries	Testicular size, sexual activity, plasma levels of FSH, LH, testosterone, prolactin		8.0–16.0	7.3	x	Lincoln (1979)
Ovis aries	Aggressive behavior wool growth, plasma levels of FSH and LH		8.0–16.0	7.3		Lincoln (1984)
Ovis aries	Estrous, plasma LH		8.0–16.0	5.9	x	Legan and Karsch (1979, 1983)
Ovis aries	Estrous		8.0–16.0	5.9	x	Legan and Winans (1981)
				7.9	x	
Ovis aries	Serum progesterone, prolactin		8.0–16.0	12.0	x	Kennaway et al. (1983)
				6.0	x	
Ovis aries	Testicular size, frequency of LH-pulses		8.0–16.0	6.0	x	Lindsay et al. (1984)

[a] Zugunruhe and molt with a 1:1 ratio, body weight with a 1:2 ratio.
[b] Natural lighting conditions.
[c] 3 anaimals full, 2 animals 2:1.
[d] Interval between every second step.
[e] Step size: 2 h.

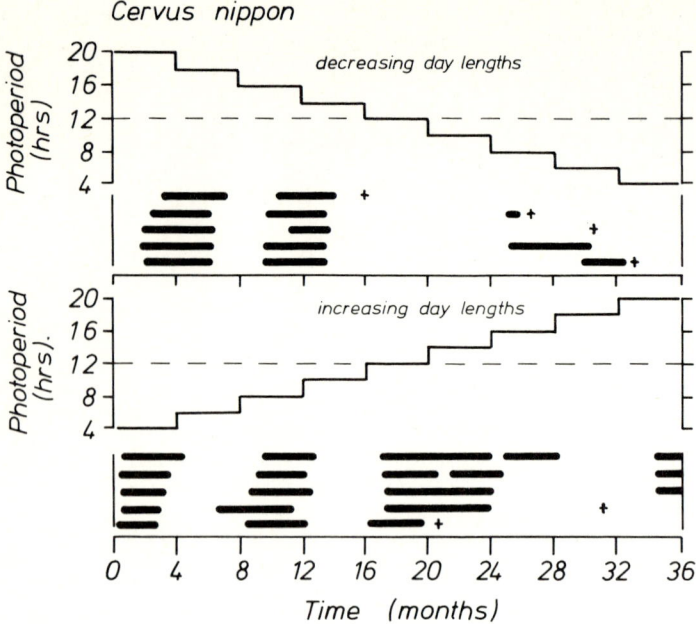

Fig. 4.5. Phases of antler growth (*bars*) in individual sika deer (*Cervus nippon*) exposed to stepwise decreasing (*upper diagram*) and increasing (*lower diagram*) photoperiods (*solid lines*). (*Crosses* indicate death of animal. (After Goss 1976)

ing to protocol 2 (p. 50). Phase-shifting of the environmental rhythms (protocol 3, p. 50) on the other hand, has been employed only rarely. In the example shown in Fig. 4.6, a group of ewes (*Ovis aries*) was first kept under the natural photoperiodic conditions and then switched to a sinusoidal photoperiodic cycle 180° out of phase with the natural one. The circannual estrous rhythms of these animals followed the phase shift almost instantaneously, despite the fact that the animals were still exposed to the original variations of environmental temperatures. Similar results have been obtained in other phase-shifting experiments with sheep (e.g., Wodzicka-Thomaszewska et al. 1967; Pelletier 1973); sika deer (Goss 1969b, 1980); common dormice (*Glis glis:*; Morrison 1964); carpet beetles (*Anthrenus verbasci:* Blake 1960, 1963); and snails (*Helix aspera:* Bailey 1981).

The results of a few experiments from which phase-dependent shifts of a circannual rhythmicity after treatment with photoperiodic stimuli (protocol 4, p. 50) can be inferred are also consistent with the role of photoperiod as a circannual zeitgeber. Figure 4.7 shows that the testicular response of starlings to continuous light and darkness depended on the time of year of transfer to constant conditions. The effect of constant light was, by and large, opposite to that of constant darkness. Although conclusive evidence is lacking, this differential testicular response to photoperiod might reflect different phase shifts of a circannual rhythmicity, which would then be consistent with the conclusion derived from other experiments that photoperiod is a zeitgeber for circannual rhythms of starlings. Phase-dependent shifts of circannual functions to photoperiodic stimuli have also

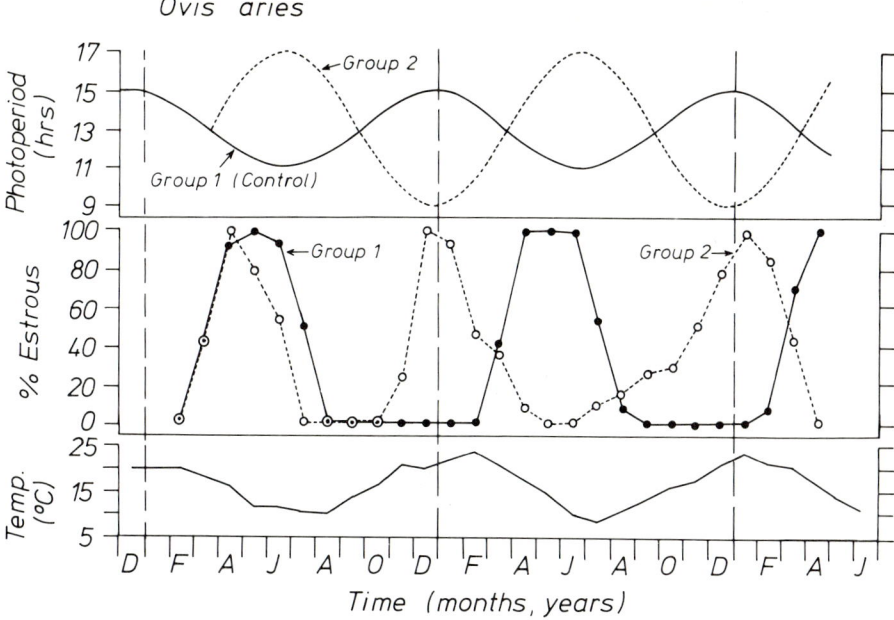

Fig. 4.6. Estrous activity (expressed as percentage of individuals in estrous) in two groups of Southdown ewes (*Ovis aries*) exposed to natural photoperiodic variations of 30°30'S (*solid lines*) or to an artificial photoperiodic cycle of a larger amplitude, 180° out of phase with the natural one. Temperature variations to which both groups were exposed are indicated on the *lower diagram*. (After Thwaites 1965)

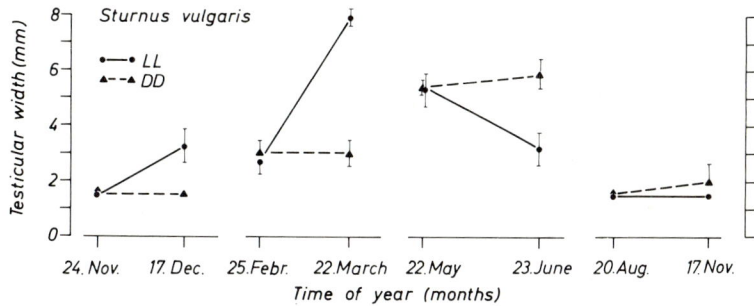

Fig. 4.7. Changes in testicular width in eight groups of European starlings (*Sturnus vulgaris*) transferred at four times of the year to constant darkness (DD: ▲) or constant light of about 0.7 lx (LL: ●). Testicular width was measured at the beginning of the experiment and again approximately 1 month later. *Vertical lines* indicate standard errors. (After Gwinner 1973)

been shown for other species, e.g., sylviinine warblers (see Gwinner 1971a for detailed discussion).

Among the thoroughly studied species there are only very few for which photoperiod appears to be ineffective as a circannual zeitgeber [although the evidence is often rather indirect; e.g., in the carpet beetle, Blake (1960, 1963)]. In ground squirrels various experiments involving photoperiodic manipulation have proved rather negative (Pengelley and Fisher 1963; Morrison 1964; Pengelley and Asmundson 1974). However, mainly short-term studies have been carried out so far with these species. In view of the slow response of some circannual systems to zeitgeber stimuli (e.g., Fig. 4.1), clear-cut responses might only be found in long-term experiments. Even then, the difficulties may be formidable, as illustrated by the experiments performed by Davis and Swade (1983) on California ground squirrels (*Spermophilus beecheyi*). These authors exposed squirrels to a 12-month rectangular photoperiodic cycle in which 6 months of LD 14:10 alternated with 6 months of LD 10:14. It turned out that even after several years the phases of the animals' body weight and molt rhythms changed systematically relative to the photoperiodic cycle. This could either indicate that the rhythms cannot be entrained by such cycles, or that they can be entrained, but that it takes a long time for them to reach a steady-state phase relationship to the zeitgeber. When squirrels were exposed to sinusoidal photoperiodic variations simulating those of their home area, the body weight rhythm of the only animal for which data are given assumed a period close to 1 year, but as 1 year is within the range of free-running period values measured in constant conditions (Table 2.1), this result cannot be considered conclusive evidence for synchronization. Phase-shift experiments also failed to provide such evidence. In a ground squirrel exposed from an age of 5 months to a sinusoidal photoperiodic cycle that was phase-shifted by 6 months, the body weight cycle kept changing its phase relative to the photoperiodic cycle throughout the 6-year experiment. Once again, it is not clear whether this indicates the free-running of the rhythm, or its transient behavior preceding steady-state entrainment.

4.1.2 Ambient Temperature

To date no convincing direct demonstrations of cycles in ambient temperature as zeitgebers for circannual rhythms are available. However, there are data suggesting that ambient temperature might act as a zeitgeber for some insects and mammals. In the carpet beetle, for instance, maxima in the circannual pupation cycle occurred at the same time and with the same precision in two groups of larvae exposed to the same natural temperature fluctuations, regardless of whether the animals were exposed to the normal photoperiodic variations or not. In contrast, maximum number of pupations occurred at another time in a group held under constant conditions of both light and temperature (Blake 1960). In the golden-mantled ground squirrel (*Spermophilus lateralis*), Pengelley and Fisher (1963) provided the first strong evidence that temperature can phase-shift the circannual rhythm of hibernation (Fig. 4.8). Seven groups of ground squirrels were held in an ambient temperature of 35 °C for different lengths of time from August, and then transferred to an ambient temperature of 0 °C. It can be seen that the high temperature prevented all animals from hibernating. After transfer to low temperature, the animals of groups 1 to 4 immediately began hibernation, but all arose at the same time in April and May, and the subsequent hibernation periods were not shifted. However, as the duration of high temperature exposure in-

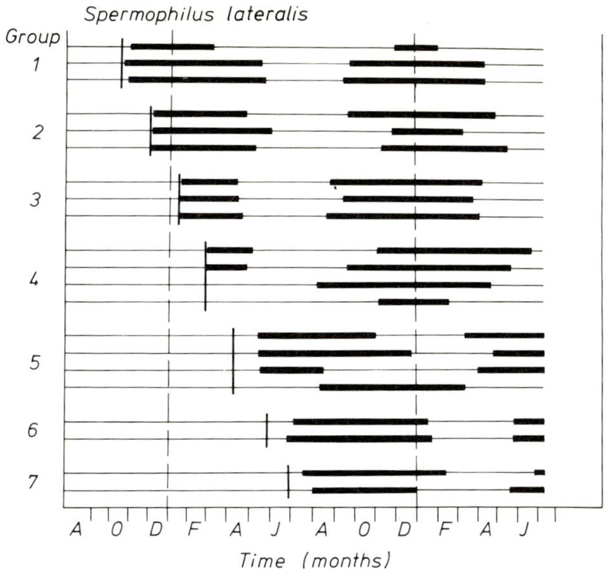

Fig. 4.8. Hibernating periods of seven groups (two to three individuals each) of golden-mantled ground squirrels (*Spermophilus lateralis*) held for 2 years in a constant LD 12:12. All animals were initially subjected to 35 °C, but at the times indicated by *vertical lines* the respective groups were transferred to 0 °C. (After Pengelley and Fisher 1963)

creased in groups 5 to 7, the arousal times and the subsequent hibernation onsets were phase-shifted by about half a year.

That temperature-induced phase shifts result in permanent shifts of the underlying circannual clock has been clearly demonstrated in another experiment with the same species (Mrosovsky 1980a) in which a control group of squirrels was held for a period of about 3 years at 21 °C and under a constant 12-h photoperiod. In these conditions the circannual rhythm in body weight free-ran with a period shorter than 12 months. In Fig. 4.9, the dates of maximal body weight in successive years are plotted underneath each other. The experimental group was held under the same conditions except for 9 months from December of the first to September of the second year, when temperature was decreased to -3 °C. This exposure to low temperature resulted in a delay phase shift of 4.5 months in the rhythm of body weight which was maintained in the following cycle. The reproductive rhythm, estimated by scrotal pigmentation and descent of testes was shifted to a similar degree. These findings, as well as those of Pengelley and Fisher mentioned above, are consistent with the hypothesis that there is a phase-response curve to temperature and, hence, that temperature cycles might function as zeitgebers for some circannual rhythms. Since the phase delay observed after a 9-month exposure to low temperature in spring and summer (Fig. 4.9) was similar to the lengthening of the period produced by continuous cold (Mrosovsky 1980b), these data also suggest that only the spring and summer phases of the endogenous circannual cycle are susceptive to synchronization by ambient temperature. Thus, there may be a long "dead zone" in the response curve during which ambient temperature has no effect. More recent data obtained by Joy and Mrosovsky (1983, 1985) from thirteen-lined ground squirrels (*Spermophilus tridecemlineatus*) are consistent with this hypothesis. Heller and Poulson (1970) on the basis of their study with ground squirrels and chipmunks also came to the con-

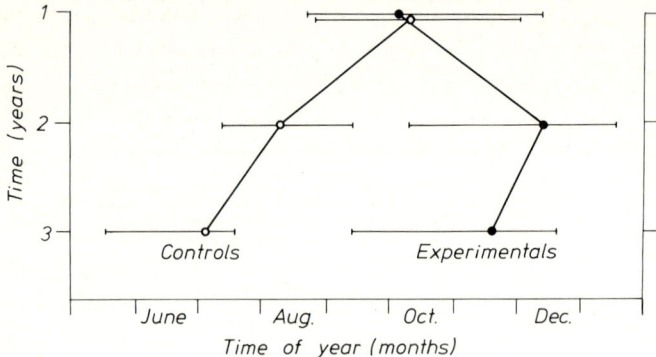

Fig. 4.9. Medians (with ranges) of maximum body weights in two groups of golden-mantled ground squirrels (*Spermophilus lateralis*) held for nearly 3 years in LD 12:12. The control group was held at 21 °C throughout the experiment, whereas the experimental group was exposed at the beginning of the experiment from December of the first to September of the second year to −3 °C before they were returned and subsequently held to the end of the experiment at 21 °C. (After Mrosovsky 1980a)

clusion that synchronization results mainly from environmental influences during the spring phase.

Apart from that of ground squirrels, there is evidence that the cycle in ambient temperature may act as a zeitgeber in European hamsters (*Cricetus cricetus*: Canguilhem et al. 1973) and common dormice (Jallageas and Assenmacher 1984). When held under a constant photoperiod, the annual rhythms of body weight of both species and the rhythms of hibernation, plasma levels of testosterone and thyroxin in dormice were more clearly expressed and more synchronous among individuals under naturally changing temperatures than in constant temperature conditions. Thus far, the common dormouse seems to be the only species in which both the annual cycle of photoperiod (cf. p. 56) and that of ambient temperature may be circannual zeitgebers.

4.1.3 Social Stimuli

The possibility that social cues may be involved in synchronizing circannual rhythms has barely been explored. However, results obtained on sheep suggest that stimuli emitted by rams can synchronize the circannual reproductive rhythm of ewes (Legan and Karsch 1983). In these experiments it was shown that the circannual rhythms of plasma LH in estradiol-treated ovariectomized ewes remained synchronized with a 6-month photoperiodic cycle when ewes were blinded but held together with a sighted ram. Once the ram was removed from the room, the 6-month rhythmicity disappeared and a longer circannual period became evident. A possible interpretation of these data is that blinding abolished photoperiodic entrainment of the rhythms of the ewes. The sighted ram, in contrast, remained synchronized with the photoperiodic cycle and transmitted photoperiodic information to the blinded ewes in a way as yet unknown. The question remains, however, whether true synchronization or only some kind of "masking" was achieved by the ram. The shape of the 6-month LH-cycles of the ewes was rather different after blinding than before. Particularly, LH-levels never decreased to the low minimal values characteristic of the time before blinding and after removal of the ram, suggesting that the ewes may not have become refractory.

4.2 Ranges of Entrainment

Since circannual rhythms persist under constant conditions for a number of cycles, they behave like technical oscillators with a certain degree of self-sustainment. One of the properties of such oscillators is that the range of zeitgeber periods to which they can be entrained is limited. This phenomenon has been amply demonstrated for endogenous circadian rhythms in which the range of entrainment normally extends from about 18 to 30 h, i.e., from a period about 25% shorter than 24 h to a period about 25% longer than 24 h. The precise values depend on the natural period τ_n of the rhythm, as well as on the strength of the zeitgeber (e.g., Aschoff and Pohl 1978; Aschoff 1980, see Appendix).

For circannual systems, the ranges of entrainment are obviously much larger than those of circadian rhythms, as can be seen in Figs. 4.3 and 4.4 and Table 4.1. Even reductions of the zeitgeber period by 50% or more are usually followed by corresponding reductions of the circannual period.

So far, the full range of entrainment has only been determined in sika deer (Fig. 4.4). Although the circannual rhythm of antler development could be entrained by photoperiodic cycles with periods of $T=12$, $T=6$, and $T=4$ months, one cycle was omitted under $T=3$ months, and under $T=2$ months, the deer replaced their antlers only every sixth cycle, falling back to a rhythm with a period of about 1 year. At the other end of the range of entrainment, a photoperiodic cycle with a period of $T=24$ months was followed by the rhythms in three out of five animals, whereas the other two replaced their antlers every 12 months (i.e., twice a cycle), thereby indicating entrainment to a submultiple of the photoperiodic rhythm. Hence, these results indicate a highly asymmetrical range of entrainment (see Appendix) extending from about $T=4$ months to about $T=24$ months.

In the European starling, the range of entrainment may be even larger, as suggested by the fact that photoperiodic cycles as short as 1.7 or even only 1.5 months are apparently still capable of synchronizing the circannual rhythms of testes size and molt (Fig. 4.3). However, because testicular rhythms required several cycles to develop under $T=1.7$ and $T=1.5$, and because one molt was omitted under $T=1.5$, these $T=$ values may be near the lower limit of the range of entrainment.

It must be emphasized that the behavior of the starlings under the short photoperiodic cycles of T = 2.0, T = 1.7, and T = 1.5 is difficult to interpret for several reasons. Figure 4.3 suggests that the successive testicular maxima and minima of the birds of all three groups were periodically modulated. The period of this superimposed rhythmicity was about 7 to 9 months. This observation could be explained by the hypothesis that the circannual rhythmicity was not truly synchronized by these extremely short photoperiodic cycles, but rather free-ran with a period of 7 to 9 months and that the short variations in testicular size were due to more peripheral direct effects of photoperiod on the system controlling testicular size. This interpretation is unlikely, however, since in a few starlings maintained under T = 2.0 for an extended period of time, the initial amplitude modulation disappeared in the course of the experiment (Fig. 4.10). This observation, argues against the above hypothesis and suggests rather that the observed phenomenon may represent some kind of transient behavior of the circannual system following the initial exposure to photoperiodic treatment. Alternatively it may be that the observed amplitude modulation was an artifact of the large interval between successive measurements. The phenomenon might have disappeared, had the sampling intervals been decreased.

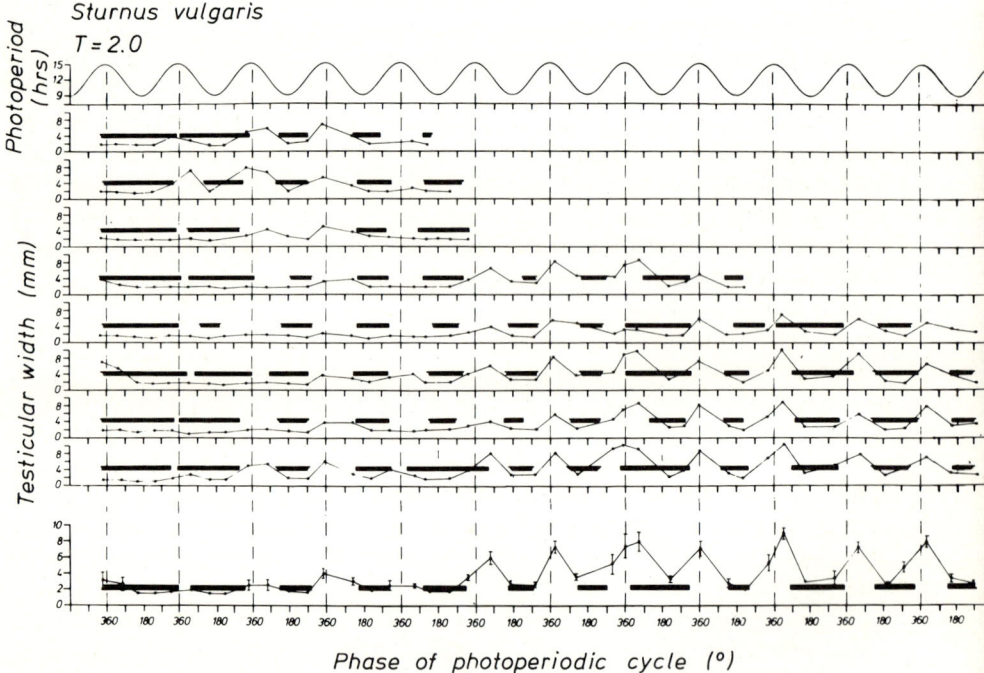

Fig. 4.10. Variations in testicular width (*curves*) and occurrence of molt (*bars*) in eight European starlings (*Sturnus vulgaris*) exposed to sinusoidal changes in photoperiod with a period of T = 2 months (*uppermost diagram*). In the *lowermost diagram* the mean values for the lower five birds are given. *Vertical lines at the symbols* and *horizontal lines at the bars* indicate standard deviations. (After Gwinner 1981b)

Another question is whether the behavior of the rhythmic functions under very short photoperiodic cycles reflects the same changes in the neuroendocrine system that occur under longer photoperiodic cycles. Several observations suggest that this may not be the case. For instance, postnuptial molt, which normally includes the whole plumage, was often reduced to a partial body molt under the very short photoperiodic cycles with periods $T = 2.0$, $T = 1.7$, and $T = 1.5$. If wing and tail feathers were renewed, the normal sequence of molt was often disturbed and incomplete. The rapid alternation between increasing and decreasing photoperiods apparently did not leave enough time for the birds to progress through the entire sequence of feather shedding.

A final aspect of the behavior of the birds living under the extremely short photoperiodic cycles which differed from the normal situation is that the testes often did not regress to the same low levels as in birds living under longer photoperiodic cycles. This was particularly so in birds which skipped their molt or reduced it to the body plumage (e.g., under a cycle of $T = 1.7$). Their beaks usually stayed yellow despite the reduction of their gonads, suggesting that testosterone was still being released at a high rate. It seems doubtful, therefore, that these birds ever became refractory in the proper sense. Hence, they presumably did not go through the whole sequence of neurendocrine stages typical for a normal bird.

In spite of these reservations, the data presented above indicate different ranges of entrainment in sika deer and starlings. Remarkably, there are even results suggesting different ranges of entrainment for different functions within one organism. In the European starling, for instance, the cycles of testicular width and molt were clearly entrained to a photoperiodic zeitgeber of $T = 3$ months or even less, but the rhythm in body weight became highly irregular in zeitgebers with periods of less than $T = 4$ months. The upper range of entrainment for the body weight rhythm may then end at period values closer to 12 months than for the rhythms of testicular size and molt (Gwinner unpubl.; Aschoff 1980). This interpretation is consistent with the idea that different functions may be controlled by independent circannual rhythms (Chap. 5).

4.3 Behavior Within the Range of Entrainment

According to a general rule in oscillator theory the phase-angle difference between an oscillator and its zeitgeber (Ψ) changes as a function of the zeitgeber period T such that Ψ decreases with a decrease in T (see Appendix). This rule, which has been amply confirmed for entrained circadian systems (Aschoff and Pohl 1978; Aschoff 1980), applies to circannual rhythms as well. As can be seen in Fig. 4.3 the phases of the starlings' circannual rhythms became progressively delayed relative to the photoperiodic cycle as the period of the photoperiodic cycle decreased. This is particularly obvious in the right diagram, in which the data are plotted as a function of relative zeitgeber time; that is, the data have been normalized in relation to the period of the photoperiodic cycle such that the zeitgeber period represents 360°, regardless of its absolute value. It is clear that with de-

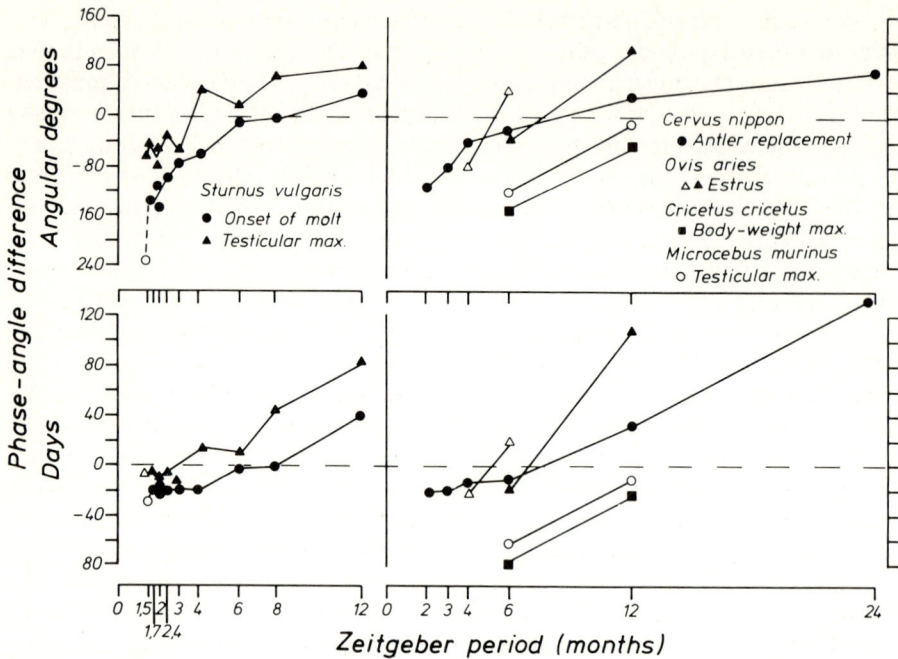

Fig. 4.11. Changes in the phase-angle ψ between circannual rhythms and photoperiodic zeitgebers as a function of zeitgeber period T ("phase curve"). On the *ordinate, positive values* indicate a lead, *negative values* a lag of the phase of the biological rhythm relative to that of the zeitgeber. In the *upper two diagrams* ψ is expressed in angular degrees, in the *lower two diagrams,* in days. The 6-month photoperiodic cycles to which *Microcebus murinus* and *Cricetus cricetus* were exposed had a saw-toothed shape, the others were sinusoidal (Table 4.1). For the studies with European starling (*Sturnus vulgaris*) and sika deer (*Cervus nippon*), the time of the longest photoperiod was used as the phase reference point, for the studies with sheep (*Ovis aries*) European hamster (*Cricetus cricetus*) and mouse lemur (*Microcebus murinus*) the time of the shortest photoperiod. References: *Sturnus vulgaris:* Gwinner 1977a, 1981b; *Cervus nippon:* Goss 1969a; *Ovis aries*: Rougeot 1961, 1962; *Cricetus cricetus:* Canguilhem et al. 1973; *Microcebus murinus:* Petter-Rousseaux 1972. (After Gwinner 1981b)

creasing zeitgeber period the maxima of testicular size as well as the onset of molt moved to later phases of the zeitgeber cycle, indicating a steady decrease of the phase-angle difference between the exogenous and the endogenous rhythms. A similar dependence is also indicated by the data obtained by Goss on photoperiodic synchronization of the circannual rhythm of antler development in sika deer (Fig. 4.4).

The dependence of Ψ on T is shown in another kind of presentation in Fig. 4.11, for the starling (left) and the sika deer and several other mammalian species (right) (Additional data not included in the graph but showing the same general tendency are available, e.g., for sheep by Mauleon and Rougeot 1962; Wodzicka-Tomaszewska et al. 1967; Lincoln 1979, 1984; Legan and Karsch 1979; Legan and Winans 1981; Kennaway et al. 1983). In the upper two diagrams, the phase-angle difference is given in angular degrees, in the lower diagram in days. It is clear that in all cases the phase-angle difference increases as the period of the photoperiodic zeitgeber increases, the rate of increase being about the same in all instances.

Two features of the "phase curves" shown in Fig. 4.11 merit special attention. One is that the slope of the phase curves describing the dependence of Ψ on T is relatively small, compared with circadian rhythms. In the circannual systems studied a change of the zeitgeber period by one month resulted in a change of Ψ by about 12°. This corresponds to a 6-degree change of Ψ per 15-degree change in zeitgeber period. In circadian systems, Ψ changes about 20° per 15-degree change in T, the precise value depending, among other things, on the properties of the zeitgeber, e.g., its strength (see below). These differences in the steepness of the "phase curves" between circannual and circadian rhythms are consistent with a theoretical rule confirmed by Aschoff and Pohl (1978) for circadian systems. According to it "circadian rhythms change their phase-angle difference to the zeitgeber within the complete range of entrainment by a similar amount of degrees, irrespective of the width of the range." In other words, large ranges of entrainment should correspond to flatter phase curves than small ranges of entrainment. Since circannual rhythms have much larger ranges of entrainment than circadian rhythms, flatter phase curves are indeed to be expected.

Another interesting phenomenon that can be seen in the lower two graphs of Fig. 4.11 is that the curves showing the dependence of Ψ on T for both the starling and the sika deer seem to level off under the very short zeitgeber periods. In other words: under T-cycles shorter than about 4 months Ψ shows only little, if any, change. It is not clear as yet how this phenomenon should be properly explained, but one possibility is that under these short zeitgeber cycles the circannual system loses its self-sustainment and begins to behave like a passive system. Such an interpretation is not unlikely, because a loss of self-sustainment under zeitgeber periods far from 24 h has been documented for circadian systems (Kreuels et al. 1984). Another possibility is that under these short photoperiodic cycles we are not dealing with true synchronization but rather with the direct ("masking") effects of the photoperiodic stimuli which influence the control system at a level different from that under the long photoperiodic cycles. Hence, these observations nourish the suspicion (p. 61 ff.) that the behavior of the system under short photoperiodic cycles may be qualitatively different from that under the longer photoperiodic cycles.

As mentioned above, both the range of entrainment and, because of the inverse relationship between it and the slope of the "phase curve" discovered by Aschoff and Pohl (1978), the slope of the phase curve should depend on the strength of the zeitgeber (see Appendix). In general, an increase in zeitgeber strength should result in an increase in the range of entrainment and a decrease in the slope of the phase curve. In the field of circannual rhythms, no systematic long-term investigations on effects of zeitgeber strength on these parameters have been carried out. However, the data summarized in Fig. 4.12 suggest that zeitgeber strength may affect Ψ even if T is kept constant. In these experiments European starlings from Germany (about 50 °N) were held for about 16 months under photoperiodic cycles of various amplitudes, simulating conditions at various geographic latitudes within the breeding range of European starlings. It can be seen that exposure to these various photoperiodic regimes had no effect on the

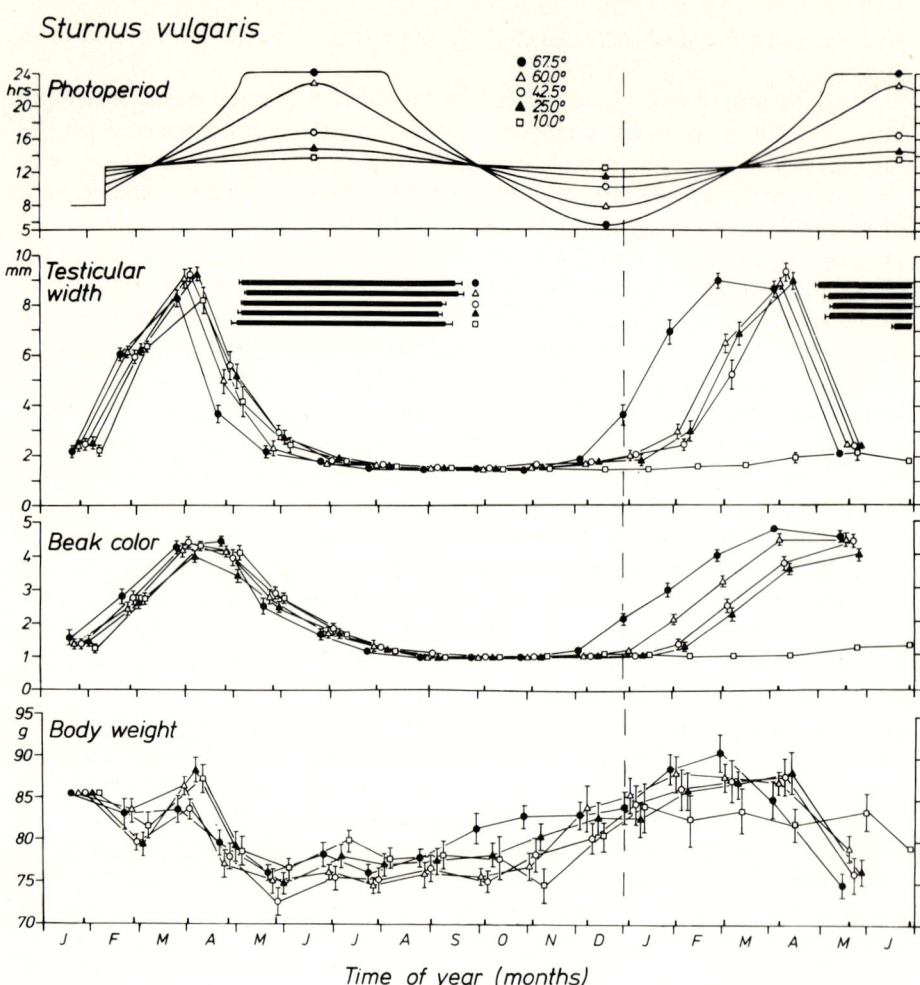

Fig. 4.12. Seasonal changes in body weight, bill coloration, and testicular width as well as occurrence of molt in five groups of European starlings (*Sturnus vulgaris*) (N = 9–12) exposed to photoperiodic cycles simulating those occurring at 5 different latitudes (*upper diagram*). A beak color index of 1 represents an entirely black, a value of 5, an entirely yellow beak. *Points* show mean values with standard errors of the means. (After Gänshirt and Gwinner 1979 and unpublished)

circannual cycles during the first 10 months of the experiment, but that differences between groups developed thereafter. In the birds exposed to the photoperiodic cycle with the largest amplitude, testicular growth began earlier than in the birds of the other groups, whereas in the birds held under a 10 °N photoperiodic cycle only marginal growth was observed, and molt was delayed. Yellowing of beaks, indicating testosterone production by the testes (Witschi and Miller 1938), began the earlier, the larger the amplitude of the photoperiodic cycle. These results might be interpreted as indicating that zeitgeber strength, defined here by its amplitude, possibly affects Ψ, but the data are not sufficient to draw any further conclusions.

4.4 Comparison with Circadian Rhythms and Some Conclusions

As outlined in the previous paragraphs, circadian and circannual rhythms share many common features with regard to their behavior in the synchronized state (Table 4.2). For both, light (photoperiod) and ambient temperature appear to be the dominating zeitgebers. Both rhythms follow phase shifts of zeitgebers through transients until the initial phase relationship with the zeitgeber is re-established. The range of zeitgeber periods to which the rhythms can be entrained is limited in circannual as in circadian rhythms. In circannual and circadian rhythms the phase relationship Ψ between biological oscillation and zeitgeber changes systematically with changing period of the zeitgeber (T) so that Ψ increases with an increasing T. At the limits of the range of entrainment synchronization by frequency demultiplication or frequency multiplication may occur in both circadian and circannual rhythms.

The major difference between the circadian and circannual systems studied so far is the much larger range of entrainment of circannual rhythms. Large ranges of entrainment are generally characteristic for oscillating systems with a low degree of self-sustainment (see Appendix). Hence these results are consistent with the observations discussed in Chap. 3 that circannual rhythms often show a tendency to dampen and are usually expressed only under a limited set of environmental conditions. The large range in period that circannual rhythms may assume under constant conditions and the relative "sloppiness" of circannual rhythms (expressed, e.g., in their large intra-individual variability, p. 42) also point to the conclusion that at least some circannual rhythms may be "weaker" systems than circadian rhythms in the sense that they are more dependent upon and more easily affected by environmental zeitgebers.

A general conclusion of this chapter is that many of the formal properties of synchronized circannual rhythms can be understood on the basis of the oscillator model of biological rhythms. This is not to say that we can expect this model eventually to accommodate all properties of all circannual rhythms. Mrosovsky (1980a) and Joy and Mrosovsky (1985), for instance, have emphasized that some of the effects of low temperature on the circannual system of ground squirrels

Table 4.2. Comparison of formal parameters of synchronization between circadian and circannual rhythms

	Circadian	Circannual
Dominating zeitgebers	Light, temperature	Photoperiod, temperature
Behavior after phase shift of zeitgeber	Follows through transients	Follows through transients
Range of entrainment	Small	Large
Behavior of Ψ as T changes	Increases at a high rate with increasing T	Increases at a slow rate with increasing T
Behavior under extremely long or short zeitgeber periods	Synchronization by frequency division	(Synchronization by frequency division)

could possibly be interpreted by assuming that low temperatures prevent the transition from one cycle to the next, i.e., arrest rhythmicity. Under natural conditions such effects could lead to synchronization, although the mechanisms responsible for it would be different from those responsible for the entrainment of a self-sustaining oscillator. Similar "entrainment-analogue" phenomena might also explain synchronization with photoperiodic cycles in species (like the European starling) that arrest rhythmicity at particular phases in most of the photoperiods normally experienced by the animals in nature. Formally, these two kinds of mechanism are clearly separable. For instance, the hypothesis of "synchronization by arresting the rhythm" predicts that the range of entrainment is unlimited towards long zeitgeber periods, since it should be possible to keep the clock arrested as long as conditions are unfavorable (e.g., Joy and Mrosovsky 1985). The entrainment hypothesis, in contrast, predicts that the range of entrainment is limited, both toward short and long zeitgeber periods (see Appendix).

Finally it must be emphasized that even in cases in which the rhythmic performance of an organism can be adequately described with the oscillator model, alternative approaches may lead to other and possibly deeper insights into particular problems. For instance, although treating circannual rhythms as entrained oscillators may help to understand their adaptive significance (Chap. 6), other approaches may be more adequate for analyzing their physiological basis (Chap. 5).

Chapter 5

Mechanisms of Circannual Organization

The present chapter is concerned with some aspects of the mechanisms that may be involved both in the generation of circannual rhythms and in the regulation of overt circannual functions by basic rhythmic processes. The chapter covers a wide and heterogenous field whose content was not easy to structure. The difficulties were further aggravated by the fact that the few investigations that have been devoted to problems of circannual physiology have been carried out on quite different species, whose circannual systems may be physiologically rather different. The available data have been organized into three sections. In the first (5.1), possible interactions between circannual and circadian rhythms are discussed. In the second section (5.2) the question of the dependence or independence of various circannual functions within an organism is investigated. The final section (5.3) reports on attempts to identify components of specific circannual rhythms.

5.1 Interactions with the Circadian System

For a variety of organisms there is evidence that a circadian rhythmicity is intimately involved in the mechanism of photoperiodic time measurement and thereby in the process of synchronization of circannual rhythms with the natural year (Bünning 1936; for reviews see, e.g., Gwinner 1975c, 1981a; Hoffmann 1981). In some species the persistence of a circannual rhythmicity in constant conditions depends on photoperiod and in turn on a circadian clock that measures it (Schwab 1971; Gwinner and Eriksson 1977; Schleußner 1984; Gwinner et al. 1985b). Moreover, it has been found that the states of the circadian and the circannual systems within an organism may show concurrent variations (see below). Observations of this kind have led to the belief that circadian rhythms may in some way be involved in the production of circannual rhythms. Circadian rhythms are widely distributed throughout the living world and almost certainly phylogenetically older than circannual rhythms (Farner 1970) so they would have been available to organisms when they evolved their endogenous annual systems. Three different kinds of hypotheses about the involvement of circadian rhythms in the production of circannual rhythms are presented. The first, discussed in Sect. 5.1.1, proposes a mechanism that may actually generate circannual rhythmicities, in essence, by counting circadian days. The other two, presented in Sect. 5.1.2 and 5.1.3, suggest that circadian mechanisms somehow convey circannual information to subordinate parts of the system, leaving open the question of how the circannual rhythmicity is generated (see also Gwinner 1975c, 1981c, Mrosovsky 1978).

5.1.1 Frequency Demultiplication of Circadian Rhythms

In attempting to develop hypotheses about physiological mechanisms underlying circannual rhythms, the most troublesome feature is the long-term nature of the processes involved. In an attempt to overcome these difficulties it has been suggested that circannual rhythms might be generated by rhythms with higher frequencies, e.g., circadian rhythms, in a process called frequency demultiplication or frequency division in oscillator technology. A well-known example of this is the electrical clock that transforms the frequency of 50 or 60 cycles per sec of the commercial electrical current into a 1 cycle per day frequency to provide information about the time of day. In an analogous way organisms may be capable of transforming circadian frequencies into a circannual frequency to provide information about the time of year. In effect such a mechanism would amount to the organism's counting subjective circadian days, "knowing" that a year has about 365 days.

This frequency demultiplication hypothesis (FDH) has been proposed more or less explicitly several times (Gwinner 1973, Farner and Follett 1979, Sansum and King 1976) but only rarely has it been tested directly. Still, there are three sets of data that argue against such a mechanism, as will be shown in the following sections (5.1.1.1–5.1.1.3).

5.1.1.1 Relationship Between Circadian and Circannual Period Length

The FDH predicts that the period of the circannual rhythmicity depends on the period of the underlying circadian rhythmicity such that the periods of the two rhythms should be proportional to one another. A preliminary test of this hypothesis (Gwinner 1973) was based on the fact that in conditions of constant light, the period of both circadian and circannual rhythms is subject to considerable individual variation. The FDH would predict that the longer the average circadian period of a bird, the longer its circannual period should be. Results obtained from nine European starlings (*Sturnus vulgaris*), whose circadian rhythms of locomotor activity and circannual rhythms of molt were determined in continous dim light, were partly consistent with this prediction. Figure 5.1 shows that there was a weak positive correlation ($p < 0.05$, Spearman's rank correlation) between circadian and circannual periods. However, questions remain about whether these results really support the FDH. First, circadian period length could not be determined in all birds throughout the experiment, since the birds tended to show arrhythmic activity patterns at times at which their testes were enlarged (see Sect. 5.1.3.1 and Fig. 5.6). Therefore, the average circadian period calculated may not be representative. Secondly, the hypothesis would have predicted a 1:1 relationship between the circadian and the circannual periods, i.e., the slope of the regression line in Fig. 5.1 should have been 1 and not 2.4 as was the case. Additional assumptions would have to be made to reconcile this deviation with the FDH.

A more rigorous test of the model was performed with male starlings and garden warblers (*Sylvia borin*) exposed to light-dark cycles with periods of $T = 22$, $T = 24$, or (for the garden warblers only) $T = 26$ h for up to 43 months (Gwinner 1981a). Each LD cycle consisted of equal durations of light and darkness (i.e., LD

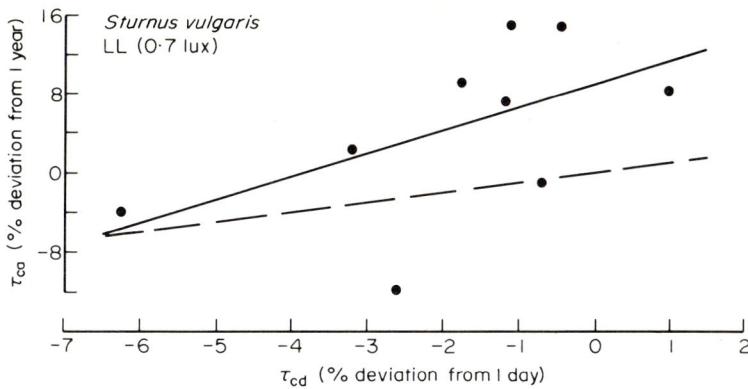

Fig. 5.1. Correlation between the average circadian period τ_{cd} of locomotor activity and the circannual period τ_{ca} of molt in nine European starlings (*Sturnus vulgaris*) kept for 15 months in constant dim light of 0.7 lx. The observed periods are expressed as percent deviation from 24 h or 365 days, respectively. *Dashed line* y = x; *solid line* regression computed from actual data, y = 2.4 × +88. (Gwinner 1973)

11:11; LD 12:12; LD 13:13). Recordings of locomotor activity indicated that the circadian perch-hopping rhythms of these birds were synchronized with all three regimes. The periods of the circannual rhythms of molt and testicular size were measured in the starlings, in the garden warblers only those of testicular size.

The results obtained from the starlings were in conflict with the hypothesis. Birds held in the LD 12:12 cycle tended to have shorter circannual periods than those in LD 11:11. Similarly, the garden warbler results disagreed with the prediction: according to the model, the birds in LD 11:11 should have had the shortest circannual period and those in LD 13:13 the longest. In contrast with this prediction, the birds in LD 12:12 had the longest period, those in LD 11:11 had the shortest and those in LD 13:13 were intermediate (Fig. 5.2). Hence these results did not support the FDH. Still, several objections could be put forward to rejecting the FDH on the basis of these results:

a) It might be argued that although the circadian rhythms of locomotor activity were synchronized with the light-dark cycles, other circadian rhythms including those involved in the proposed counting mechanism were not. Such a possibility cannot be excluded with certainty. It seems unlikely, however, since many other studies have indicated that light-dark cycles with periods of T = 22, T = 24, and T = 26 h, and with the amplitude used in the present experiment are well within the range of entrainment of circadian functions in birds and other vertebrates (Aschoff and Pohl 1978).

b) A second possible objection is that the data represent only the behavior of circannual rhythms during the first cycle in constant conditions when the system was probably not yet in steady state (Chap. 3.2), and that the predicted relationship between circadian and circannual rhythmicity would only have been found in subsequent cycles. This objection is not very relevant, however, as any useful FDH would have to explain the transient behavior of circannual rhythms as well. In fact, as will be pointed out later (Sect. 5.1.1.3), the existence of transients provides a major a priori argument against such a mechanism.

c) A third possible objection to the rejection of the FDH might be derived from the fact that in these experiments the period of the light-dark cycle was altered by altering both the light and the dark fraction of the cycle. A change in cycle length, therefore, also results in a change of the absolute duration of the light time. This complicates the issue, as starlings and garden warblers are known to react differently to different constant photoperiods (e.g., Sect. 5.3.2). On top

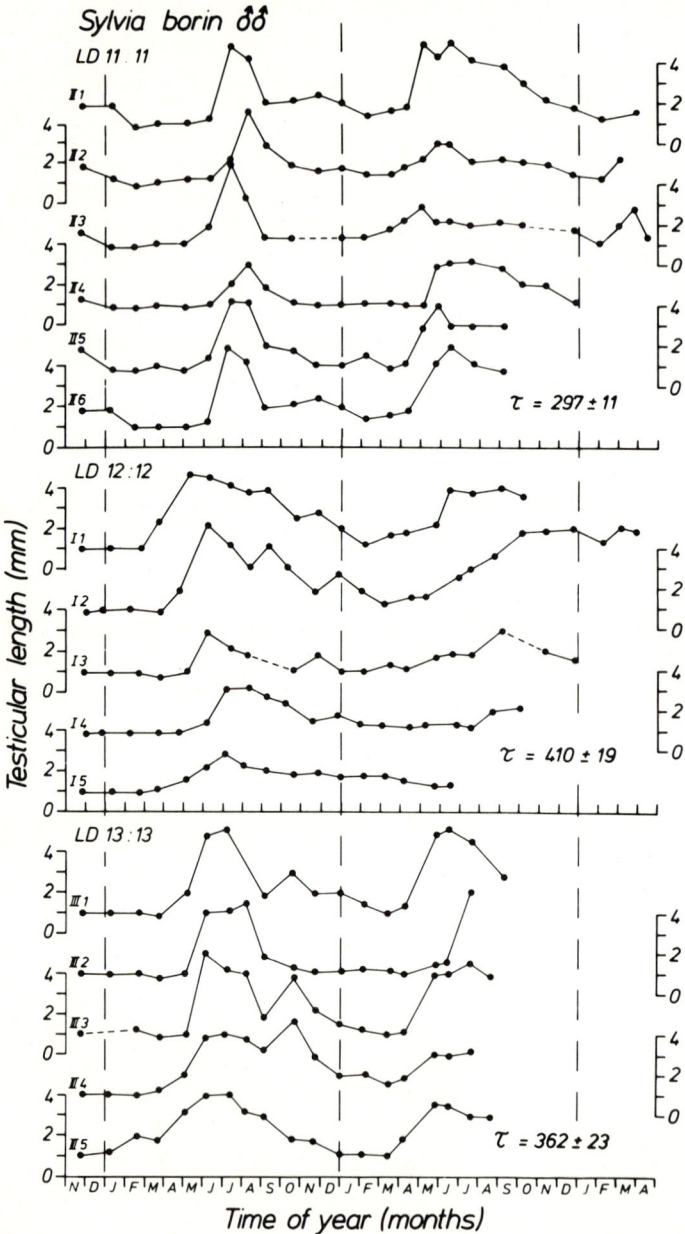

of this, the phase relationships between circadian rhythms and the entraining light-dark cycle change as a function of the zeitgeber period (Aschoff 1980; Aschoff and Pohl 1978). In a 22-h cycle circadian rhythms assume a more negative phase relationship to the zeitgeber than in a 24-h or 26-h cycle (see Appendix), so additional photoperiodic effects may have resulted from these different phase relationships. It could be argued, therefore, that circannual rhythms still result from frequency demultiplication, but that the effects of this mechanism are obscured or superimposed by photoperiodic effects. Once again, such a possibility, together with the fact that circannual cycles can indeed be heavily affected by photoperiod, must be taken as an argument against the FDH rather than as an argument against its rejection (see Sect. 5.1.1.3).

Some results obtained from studies with mammals are also in conflict with the FDH. In four antelope ground squirrels (*Ammospermophilus leucurus*) held for 19 months in LL there was no correlation between the period of their circannual testicular cycle and the average period of their circadian rhythm of locomotor activity (Kenagy 1981b). However, as in starlings (see above) circadian period could not always be determined, so that the "average" circadian periods may not be exactly representative. Richter (1978) recorded locomotor activity, food and water intake, and body weight changes of a blinded chipmunk (*Tamias striatus*) held in LD 12:12 for 6.5 years (see Fig. 2.3). All four functions showed a remarkably constant circannual rhythmicity, in which successive periods varied only between 320 and 340 days. In contrast, the period of the circadian activity rhythm was quite variable; in some years it averaged less than 24 h, in others more than 24 h; also, over long periods of time no circadian patterns were discernible at all. Richter concluded that these findings "rule out the possibility that changes of the 24-h clock could have determined lengths of the yearly periods."

Together, almost all the results presented here argue against the existence of a simple frequency demultiplication device as the basis for circannual rhythms. Only the weak positive correlation found between circadian and circannual period length of starlings held in continuous dim light (Fig. 5.1) is consistent with it. However, in view of the negative evidence obtained from the experiments with starlings whose circadian rhythms were driven by light-dark cycles with different periods, it seems likely that this correlation resulted from the action of common physiological factors affecting both rhythms, rather than from a causal relationship between these two rhythmic processes.

Fig. 5.2. Variations in testicular length in male garden warblers (*Sylvia borin*) kept under the three different constant light-dark cycles given on the *upper left* of each panel. Circannual period length was defined as the interval between dates at which testicular length during the first two phases of testicular growth exceeded 2.0 mm. The τ values (measured in 24 h days) in the *lower right* of each of the three panels represent means with standard deviations. The frequency demultiplication hypothesis was tested statistically by converting the circannual period values of the individual birds from "days" into number of "light-dark cycles" passed between the first and the second onset of testicular growth (as defined above). The null hypothesis for the hypothesis then is that τ is the same in all groups. This, however, was only true for the comparison of the group held in LD 11:11 with that held in LD 13:13 but not for the other two comparisons. (Gwinner 1981a)

5.1.1.2 Effects of Disrupting the Circadian System on Circannual Rhythmicity

If organisms derived circannual rhythms via frequency demultiplication from their circadian system, the abolition or impairment of the latter should result in the abolition or impairment of the former. The following data are relevant in this context.

a) In the European starling, as in other passerine birds (for review see Menaker and Binkley 1981), pinealectomy leads to a severe disruption of circadian locomotor activity rhythms under conditions of continuous dim light or darkness. The effects vary from the rhythmicity becoming highly unstable and subject to frequent changes in period length, to a complete loss of rhythmicity (Gwinner

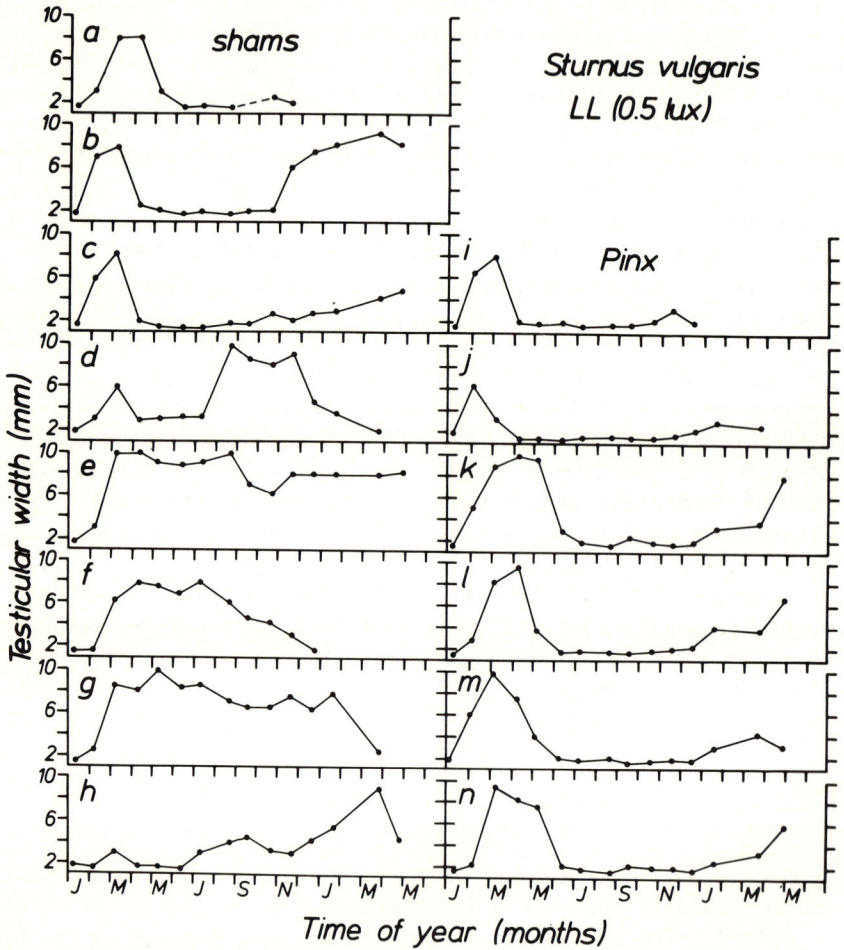

Fig. 5.3. Variations in testicular width in individual sham-operated (*shams*) and pinealectomized (*Pinx*) European starlings (*Sturnus vulgaris*) kept for up to 17 months in continuous dim light of about 0.5 lx. (Gwinner et al. 1981)

1978). Still, a circannual rhythm in testicular size persisted in six pinealectomized starlings held for 16 months in continuous dim light (Fig. 5.3). In fact these birds showed even more consistent circannual rhythmicity than the eight control birds held under the same conditions (Gwinner et al. 1981).

b) In several rodent species, destruction of the suprachiasmatic nuclei (SCN) of the hypothalamus severely disrupts circadian rhythmicities in both free-running and entrained states (Rusak and Zucker 1979; Turek 1983, for reviews). Nevertheless, circannual rhythms in body weight and reproductive function have been shown to persist in some golden-mantled ground squirrels (*Spermophilus lateralis*) held in continuous light, whose circadian activity rhythms were abolished or heavily disrupted by SCN lesions (Zucker et al. 1983; Dark et al. 1985). Preliminary data of a similar kind had previously been obtained by Mrosovsky (1975).

c) In most organisms, overt locomotor activity rhythms disappear under conditions of continuous light, once light intensity exceeds a certain threshold. Among passerine birds, this threshold intensity varies between about 1 lx and 100 lx. There is evidence that this overt arrhythmia reflects the halting of the circadian clock(s) controlling locomotor activity (Binkley 1977). Nevertheless, circannual rhythms have been documented for several organisms, including birds, held under continuous bright light with intensities up to 300 lx (Table 2.1). Similarly, "infradian" body weight cycles with periods of about two to three months (Table 2.3) persisted in common dormice (*Glis glis*) held in LL of about 700 lx during which circadian organization of wheel-running was generally absent (Mrosovsky et al. 1980).

All these findings clearly do not support the FDH. On the other hand, they are not sufficient for its ultimate rejection. In all experiments mentioned above only one circadian function has been shown to be impaired by the experimental treatment, but it is by no means certain that the circadian system as a whole was disrupted. In fact, there is recent evidence suggesting that SCN lesions in rats and monkeys may not abolish all circadian rhythms (Prosser and Satinoff 1984; Fuller et al. 1981; Reppert et al. 1981), and that exposure to bright light may not render all circadian functions of European starlings arrhythmic (Gänshirt et al. 1984). These findings indicate then that the loss of one circadian function does not necessarily indicate that the whole system has become arrhythmic, leaving open the possibility that in the above-mentioned experiments some circadian rhythms could still have been available for use in a frequency demultiplication mechanism.

5.1.1.3 General Properties of Circannual Rhythms in Conflict with the Model

Apart from the experimental evidence mentioned above, circannual rhythms have several properties that are extremely difficult to reconcile with the FDH.

a) As shown in Chapt. 4, many circannual rhythms can be synchronized to photoperiodic cycles with periods considerably different from 1 year. Since in these experiments the light-dark cycle synchronizing the circadian rhythms always had a period of 24 h, one would have to assume that the "counting" mechanism of the system becomes modified according to the rate at which photoperiod oscillates, or is overridden by the photoperiodic changes. Although such a mechanism is theoretically possible, its existence is not particularly likely.

b) Entrained circannual rhythms follow phase shifts of the zeitgeber, but it may take several transient cycles before resynchronization is completed (i.e., during resynchronization the length of successive periods changes systematically with time, e.g., Fig. 4.1). Similar transients can also be observed during the first cycles after transfer to constant conditions (e.g., Figs. 2.8, 2.9, 2.12, and 3.2). Such changes in the circannual period could only be accommodated with the FDH if one assumed that the frequency demultiplication mechanism changes its properties continuously during the successive transient cycles.

c) In some animals the period of circannual rhythms is a function of light intensity and/or temperature, but sometimes quantitatively different than the effect of these variables on the period of circadian rhythms. An extreme example is found in the common dormouse in which the period of the (atypical) circannual body weight cycles observed in an LD 12:12 was much shorter in 5 °C (53 days) than in 22 °C (162 days) (Mrosovsky 1977 see Table 3.2). The period of the circadian activity rhythm in this species, in contrast, was only shortened by about 3% when environmental temperature was lowered by a corresponding amount (Pohl 1968). Finally interindividual variability of circannual periods of different individuals held under the same set of environmental conditions is usually much larger than the interindividual variability of periods of circadian rhythms (Chap. 3). Once again, a complicated weighting system for the animal's interpretation of the duration of a day would have to be assumed to reconcile such data with the FDH.

In summary, then, it can be stated that, although there are no compelling data against it, both theoretical considerations and negative experimental evidence render a simple frequency demultiplication mechanism rather unlikely for the species thus far studied. However, one cannot preclude the possibility that this conclusion may require revision as a result of future experimental research.

5.1.2 Circannual Rhythm of a Circadian Rhythm in Photosensitivity

According to the "external coincidence" model proposed by Bünning (1936) and refined by Pittendrigh (1966, 1972), photoperiodic reactions depend on the coincidence between light and a particular phase of a circadian oscillation (φ_i). Under natural conditions the annual cycles of overt functions result from this circadian phase becoming periodically exposed to light as a consequence of the seasonal variations in photoperiod. However, a periodic illumination of φ_i could also take place under constant photoperiodic conditions if one assumed that the phase relationship between the circadian rhythm and the constant light-dark cycle to which it is entrained were subject to circannual variation. A simplified version of this model is shown in Fig. 5.4 A. During the birds' subjective autumn and winter the phase relationship between circadian rhythm and light-dark cycle, Ψ_w, is such that light never falls on the photosensitive phase φ_i, indicated by the black block.

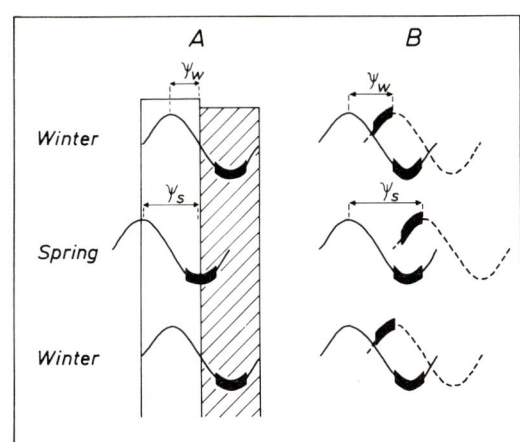

Fig. 5.4. Schematic representation of two models that derive circannual rhythms from circadian rhythms. Overt circannual rhythm resulting *A* from changes in the phase-angle difference (Ψ) between a circadian rhythm and its entraining light-dark cycle; *B* from changes in the phase-angle difference between two circadian rhythms. (After Gwinner 1973)

However, as the birds' subjective spring approaches, the circadian rhythm spontaneously changes its phase relationship to the light-dark cycle, Ψ_s, so that a fraction of φ_i becomes illuminated. As a consequence, the photoperiodic response occurs. Later on, the initial phase relationship is reestablished. Hence, in this model a circannual rhythmicity in overt functions (e.g., that of gonadal activity) results from circannual variations in the phase relationship between a circadian rhythm and its entraining light-dark cycle (e.g., King 1968; Gwinner 1973).

It should be re-emphasized, that in contrast with the frequency demultiplication hypothesis discussed in the preceding section, this model does not propose a mechanism for the generation of circannual rhythms because it leaves open the question about the factors responsible for the circannual variations in Ψ. However, the model shifts the problem to a level already partly understood, and, hence, may contribute to our understanding of the system as a whole.

There is good evidence from several species that the assumptions made by this hypothesis may be valid. The participation of a circadian oscillation in photoperiodic time measurement has been documented for several avian and mammalian species (e.g., Pittendrigh 1966; Farner and Lewis 1971; Follett 1973; Elliott 1976; Farner and Follett 1979; Turek and Campbell 1979; Hoffmann 1981) including some with circannual rhythms (Gwinner 1975c). Moreover, there are data that indicate a seasonal change in the phase relationship between circadian rhythms and the entraining light-dark cycles, apparently caused by changes in the animal's internal state (Aschoff 1969; Daan and Aschoff 1975; Turek and Gwinner 1982).

This model is not very plausible for such species that show circannual rhythms in a wide variety of different photoperiods (Tables 2.1 and 2.2). Nonetheless, it can accommodate several peculiarities of circannual rhythms in species in which the expression of a circannual rhythmicity is limited to a very narrow range of photoperiods. For instance, the circannual testicular cycle of the European starling is expressed only in LD 12:12. In longer photoperiods the rhythm stops with the gonads in a regressed state, the birds being refractory to long-day stimulation. In shorter photoperiods, in contrast, the rhythm stops with the gonads in an enlarged state (Fig. 5.12; Hamner 1971; Schwab 1971; Gwinner et al. 1985b). The long-day phenomenon can be explained by the present model by assuming that in birds kept on a photoperiod longer than 12 h, φ_i is continuously exposed to light, so that the birds never perceive the short-day stimulus required for the breaking of photorefractoriness. The short-day phenomenon can be explained by assuming that in birds kept on a photoperiod shorter than 12 h, φ_i is continuously exposed to darkness, so that the birds never receive the long-day stimulus required for the initiation of refractoriness. Only in an LD 12:12 the suggested circannual changes in the phase relationship would result in an alternate exposure of φ_i to light and darkness, whereas in shorter and longer photoperiods these changes would not be large enough to move φ_i into and out of the light.

Obvious ways of testing this hypothesis would consist in exposing animals to conditions known to affect the natural period τ_n of the circadian rhythmicity and, hence, its phase relationship with entraining light-dark cycles in a predictable way (see Appendix). Treating animals with certain drugs, e.g., D_2O or lithium chloride, which have effects on circadian period and phase (Enright 1971; Engelmann 1986) would do this, but such experiments have as yet only been carried out on a short-term basis, testing simple photoperiodic responses, and with animals which have not yet been shown to exhibit circannual cycles (e.g., golden hamsters, *Mesocricetus auratus*; Eskes and Zucker 1978). Another possibility would consist in changing the circadian system in a defined way by surgical interference. Attempts of this kind have been made in experiments with pinealectomized European starlings, whose circadian activity rhythms have different properties than those of intact birds. The complex results are difficult to interpret, possibly because the effects of pinealectomy on the circadian system are not yet sufficiently understood (Gwinner and Dittami 1980, 1982; see also p. 74).

As of yet, only evidence against this model can be cited. The most compelling results against it are probably those that demonstrate persistent circannual rhythms in conditions of continuous light or darkness (Tables 2.1 and 2.2). For such cases the present hypothesis derived from a simple external coincidence model of photoperiodic time-measurement is not applicable without considerable modification: the hypothesis assumes that the overt circannual rhythm results from an alternate exposure of a particular circadian phase to light and darkness. In continuous light (or darkness), however, any circadian phase is continuously exposed to light (or darkness). The European starling has been shown to exhibit circannual cycles in continuous dim light (Table 2.2 and Fig. 5.3). The interpretation given above (p. 77), for the persistence of a circannual rhythmicity in this species in LD 12:12 and for its arrest in LD 11:13 and LD 13:11, respectively, must therefore be taken with extreme caution.

5.1.3 Circannual Variations in the Internal Circadian System

The difficulty of explaining the continuation of circannual cycles in continuous light or darkness could be overcome if the external coincidence model were replaced by a model of "internal coincidence". This idea was first proposed by Pittendrigh 1972 (see also Tyshchenko 1966; Gwinner 1973; Pittendrigh and Daan 1976; Dolnik 1976), who suggested that light-dark cycles only entrain circadian rhythms, without simultaneously exerting a direct inductive effect as assumed by the external coincidence model. Photoperiodic reactions would then result from particular states of the circadian system, which are induced by particular photoperiods. Specifically, it has been proposed that photoperiodic alterations change the phase relationship between two or more circadian oscillators within the organisms and that a photoperiodic reaction occurs when a particular phase relationship is established. Annual changes in the phase relationship between or among oscillators may, however, not only result from environmental photoperiodic alterations, but may even take place on a circannual basis under constant photoperiodic conditions as illustrated in Fig. 5.4 B. Here the two sine curves symbolize two circadian oscillators. On each oscillator a particular phase is delimited by a black block. It is assumed (1) that the phase relationship between these two oscillators exhibits spontaneous periodic variations with the effect that the two phases coincide in spring (Ψ_s) but not at other times of the year (Ψ_w); and (2) that the photoperiodic response is initiated when these two phases are coincident. Hence in this type of model circannual rhythms in overt functions would eventually result from a circannual rhythm in the phase relationship between circadian oscillators. In contrast with the previous hypothesis, this model could explain the persistence of circannual rhythms both under light-dark cycles and constant conditions of light and darkness.

There is evidence for this hypothesis in some species. For instance, it has been demonstrated that more than one circadian oscillation exists in individual organism including higher vertebrates (e.g., Wever 1979; Aschoff 1980; Moore-Ede et al. 1982 for recent reviews). Moreover, many results show that the phase relationship between different circadian rhythms is not constant but subject to considerable variations (Aschoff and Pohl 1978; Aschoff 1980). In the following, two groups of studies relevant to this model will be discussed.

5.1.3.1 Relationship Between States of Circadian and Circannual System

The internal coincidence model predicts that there should be changes in the state of the circadian system correlated with particular phases of the circannual system and that these correlations reflect causal effects of the state of the circadian system on the circannual rhythmicity. The first prediction has received support from several studies indicating that either circadian period (τ) or circadian activity time (α) may change seasonally with the state of the gonads.

An example of a circannual cycle of circadian τ is shown in Fig. 5.5. A male golden-mantled ground squirrel held in continuous light changed the period of its circadian locomotor activity rhythm systematically with the phase of its circannual rhythm of gonadal size and body weight; the shortest periods occurred during the weight-loss phase when gonads were enlarged (Mrosovsky et al. 1976). Unfortunately, clear data are as yet available only from this one male and a few females (Zucker 1986).

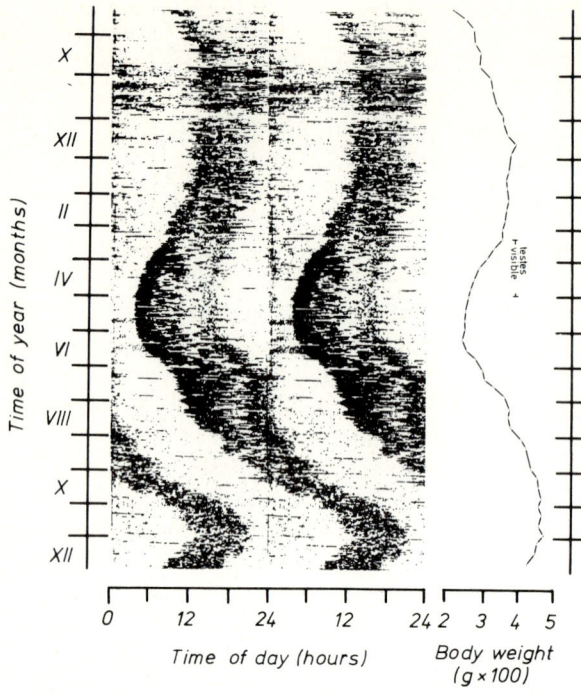

Fig. 5.5. Circadian activity rhythm (*left*) and changes in body weight (*right*) of a golden-mantled ground squirrel (*Spermophilus lateralis*) held for about 15 months in LL (25 lx). (After Mrosovsky et al. 1976)

Circannual variations in circadian activity time (α) have been demonstrated in male European starlings held in continuous light (Gwinner 1973). Generally, α tended to lengthen (and often activity time became continuous) as the testes grew. Following testicular regression – but often with a considerable delay – α shortened again (Fig. 5.6). Long-term data of this kind are available from only four starlings, but there are results from a variety of short-term studies on birds that document the same phenomenon: both in constant light and in LD-cycles α tended to be longer in birds with active gonads than in birds with inactive gonads (Hamner and Enright 1967; Aschoff 1969; Pohl 1972; Gwinner 1974a, 1975b, 1980; Rutledge 1974; Daan and Aschoff 1975). A relationship between circadian activity parameters and reproductive state has also been found in several short-term experiments with mammals (Zucker 1979; Turek and Gwinner 1982).

There is then at least suggestive evidence of changes in the circadian system associated with changes in the organism's circannual reproductive state. However, so far most of these data can be interpreted as reflecting the action of reproductive hormones on the circadian system, rather than vice versa. In male starlings, for instance, the lengthening of α that usually occurs in photosensitive birds transferred to LL or to a long photoperiod has been obliterated by castration and re-established by testosterone treatment (Gwinner 1974a, 1975b, 1980). Comparable results were obtained in experiments with mammals (Morin et al. 1977; Zucker 1979; Turek and Gwinner 1982).

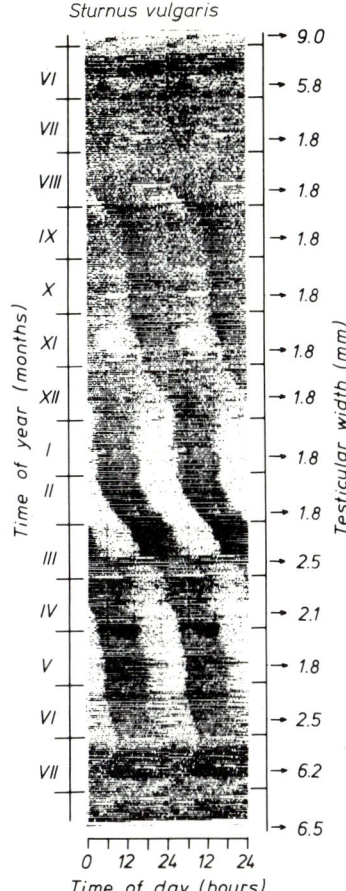

Fig. 5.6. Circadian activity rhythm of a European starling (*Sturnus vulgaris*) kept for 15 months in LL (0.7 lx). The daily activity records have been pasted underneath each other on a chart and the chart was then duplicated for better survey. *Numbers at the right-hand margin* refer to testicular width as determined by laparotomy. (After Gwinner 1973)

There are only very few data that suggest that seasonal variations in overt circadian functions may occur independently of concurrent changes in the reproductive condition. For instance, in starlings held for 15 months in continuous dim light, an increase in α occurred not only in those four birds that initiated a second testicular cycle after 8 to 13 months in constant conditions (Fig. 5.6), but also in most of the 12 birds that did not. These results, however, should not be overrated, particularly because it is possible that the latter birds went through changes in their endocrine state, despite the fact that the gonads retained their small size. Similar objections could be made in a few other similar instances (Gwinner 1973, 1980).

In summary, there is as yet no convincing evidence from the approach discussed in this paragraph, that spontaneous changes in the internal circadian state of organisms may be the direct cause of circannual physiological variations. However, none of the above-mentioned studies completely exclude such a mechanism; it is possible that the overt circadian functions studied (mainly locomotor activity) bear no relationship with those circadian functions which may be involved in the timing of seasonal activities.

5.1.3.2 Internal Coincidence Between Circadian Neurotransmitter Rhythms as the Basis of Circannual Changes?

A concrete internal coincidence model as an essential component of circannual rhythms was proposed by Meier and his coworkers for the white-throated sparrow (*Zonotrichia albicollis*) and several other vertebrate species (e.g., Meier et al. 1980; Meier and Wilson 1985). It was derived from the initial observation that the annual cycle of sparrows could be affected by appropriately timed daily injections of corticosterone and prolactin. If, for instance, photorefractory sparrows were held in LL and treated with prolactin 12 h after each daily corticosterone injection, the birds responded with some gonadal growth, vernal migratory fattening, and northward-oriented vernal zugunruhe. These vernal events did not occur if prolactin was injected 8 h after corticosterone. Consistent with these results it was found that untreated sparrows held in natural lighting conditions showed corresponding seasonal changes in the diurnal pattern of these two hormones: during the vernal migratory period the daily peaks of corticosterone and prolactin were about 12 h apart, whereas during the summer refractory period they were only about 8 h apart.

On the basis of these findings it was proposed that seasonal changes in physiological conditions normally result from changes in the temporal relationship between two circadian neuroendocrine oscillators which can be reset by corticosterone and prolactin. Since there is evidence that corticosterone stimulates the synthesis of serotonin, and prolactin that of dopamine, it was attempted to duplicate the above effects by injecting precursors of these two transmitters, 5-HTP and DOPA. It turned out that, indeed, vernal events could be induced in refractory sparrows if DOPA injections were given 12 h after 5-HTP injections: The experimental birds treated for 10 days in LL and subsequently held for about 10 months in LD 16:8 showed increased weight of gonads and of body fat stores, and intense nocturnal activity which was oriented northward. The subsequent summer and fall events also occurred in these birds: a complete postnuptial molt was carried out and followed by autumnal migratory fattening and southward-oriented zugunruhe. The control birds, in contrast, which had received only saline injections, remained in their initial photorefractory condition. These results have been interpreted to suggest that the daily injections of 5-HTP and DOPA, spaced 12 h apart, had reset the circannual mechanism. Similar results were also obtained from experiments with some other vertebrates, such as golden hamsters and Gulf kingfish (*Fundulus grandis*).

5.2 Interrelationship Among Different Circannual Functions. Or: One or Several Circannual Clocks?

The rigid temporal sequence in which various overt annual events occur, might be taken to suggest that these events or the physiological processes directly controlling them constitute integrate components of the overt circannual rhythmicity. For instance, in a migratory bird the termination of reproduction in summer, or the neuroendocrine events leading to it, might trigger postnuptial molt, the termination of which might then trigger autumnal migratory restlessness etc. Considerations of this kind have prompted some authors to ask rather explicitly whether circannual rhythms are composed of "a sequence of linked stages, each one taking a given amount of time to complete and then leading into the next with the last stage linked back into the first again" (Mrosovsky 1970; see also Enright 1970). This proposition was considered an alternative to the circannual rhythm concept for some time, but it now seems to be generally accepted that the "sequence of stages idea" proposes a methodological strategy for the analysis of circannual rhythms rather than an alternative interpretation of the phenomena (Menaker 1974; Mrosovsky 1974b). "Of course, both circadian and circannual rhythms must consist of sequences of interdependent steps" (Menaker 1974), the important question being: what steps are involved? At what levels of organization do the steps occur? And how are they interconnected?

In taking this approach, the study of circannual rhythms is in many respects easier than the investigation of other rhythmic phenomena with shorter periods, e.g., circadian rhythms. The long time constants involved in circannual cycles make it easier to analyze the various components of the system separately; indeed, some of these components are often already known or are at least reasonably open to speculation.

This section addresses the question: At what level of organization do the various circannual functions occurring within an organism interact? In the earlier literature on the subject one can find proponents of both possible extremes. Some authors favored the possibility of rather profound interactions at a rather peripheral level. Pengelley (1969), for instance, suggested that in hibernators the attainment of heavy body weights (due to fattening that normally precedes hibernation), may be "a prerequisite and perhaps a trigger to the onset of hibernation." However, the alternative view that various functions are relatively independent of one another was also suggested long ago. Blanchard and Erickson (1949), for instance, wrote about the "sequence of causation" for gonadal cycles, molt, fat deposition, and migration in birds: "It must, however, always be admitted that gonadal changes might be a separate element depending on the inherent cycle or on its own set of environmental factors while other subsequent elements might, at their own time, respond to different internal or external stimuli."

For species in which several functions have been shown to be under circannual control (e.g., in the golden-mantled ground squirrel: hibernation, body weight, food consumption, reproduction; in the garden warbler: zugunruhe, body weight, food preference, gonadal function) these considerations inevitably lead to the question of whether these various overt rhythms are part of one clock system, or whether they are basically independent of each other, i.e., expressions of separate circannual clocks. It is clear that a general answer to this question cannot be expected. Rather, different organisms are likely to behave differently and within an organism different functions may be interlinked to different degrees. The following account concentrates on cases indicating a relatively high degree of independence of various circannual functions. Evidence is derived from three kinds of data: (1) observations indicating that different circannual rhythms within an organism may persist for different periods of time (Sect. 5.2.1); (2) observations suggesting changes in the internal phase-relationships between various circannual rhythms (Sect. 5.2.2); and (3) results indicating that certain experimental manipulations can modify or abolish one circannual rhythm of an organism without affecting others (Sect. 5.2.3).

5.2.1 Differential Degrees of Persistence of Various Circannual Functions

Several investigations have indicated that under certain experimental conditions the annual rhythms of some functions continue while others damp out, suggesting a high degree of independence between these functions. For instance, in blackcaps (*Sylvia atricapilla*) held in LD 10:14, 12:12, or 16:8 circannual rhythms of zugunruhe and molt continued for up to three cycles, but the rhythm of body weight disappeared (Berthold et al. 1972a). In general there is a tendency in several passerine birds for the molt rhythm to be more persistent under constant conditions than other functions (Berthold 1974a, b). In antelope ground squirrels kept in LD 12:12 or LL, testis size showed a circannual rhythmicity while rhythms in body weight and water consumption were not discernible (Kenagy 1981b). Similarly, in thirteen-lined ground squirrels (*Spermophilus tridecemlineatus*) held in LD 12:12 a circannual rhythm in molt persisted even in those few individuals in which a rhythm in body weight was not discernible (Joy and Mrosovsky 1982). In rams (*Ovis aries*) held in an LD 6:18 a rhythm in testicular size could be observed for several cycles, but no rhythm was present in plasma prolactin; in an LD 18:6, in contrast, both rhythms continued (Howles et al. 1982). Golden-mantled ground squirrels showed a circannual rhythm of hibernation at low ambient temperatures but not at high temperatures (although a rhythm in the duration of sleep persisted; Walker et al. 1980). The body-weight rhythm, in contrast, was expressed in both conditions (Pengelley and Fisher 1963; Pengelley et al. 1978).

Even rhythmic functions normally considered to be closely linked may have a high degree of independence, as indicated by their differential tendency to dampen in a constant environment. A striking example is summarized in Fig. 5.7, which shows the time course of molt, testicular size, and bill coloration in three European starlings held for 43 months under constant conditions. The yellowing of the beak normally associated with testicular growth is considered to be the result of increased secretion of testosterone (Witschi and Miller 1938). The behavior of the bird in the upper panel represents a typical case: each time testes regressed, the beak turned black, and molt occurred. The behavior of the bird shown in the middle panel indicates that this typical relationship was not always maintained. The beak of this bird remained yellow for almost 3 years, i.e., the rhythm of bill coloration disappeared despite the fact that testicular size went through repeated cycles similar to those of the bird in the upper panel. Although alternative explanations are possible, the most likely interpretation for this dissociation is that the

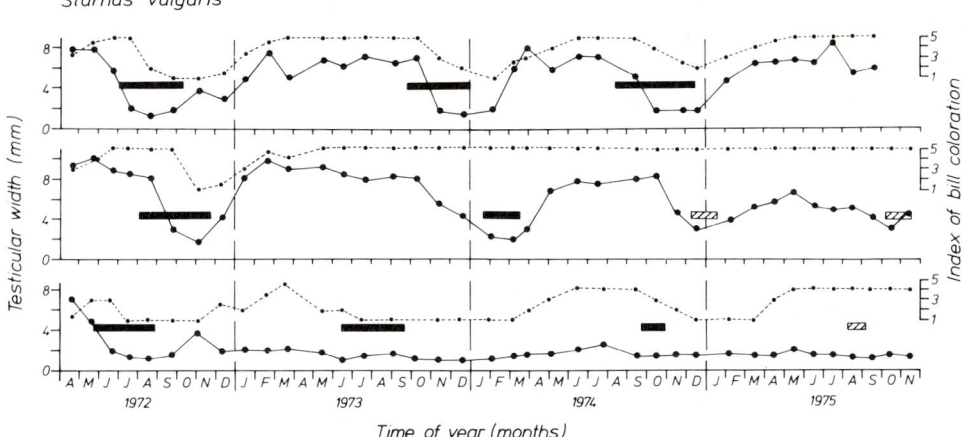

Fig. 5.7. Variations in testicular width (*solid lines*) and bill coloration (*dotted lines*) as well as the occurrence of molt (*bars*) in three European starlings (*Sturnus vulgaris*) held for 43 months in a LD 11:11 (*upper panel*) or LD 12:12 (*middle* and *lower panel*). Solid bars molt of wing and tail feathers and body plumage; hatched bars molt of body plumage only. The index of bill coloration was estimated on the basis of an index of *1–5*, 1 characterizing a black and 5 an entirely yellow beak. (After Gwinner 1981a)

spermatogenetic tissue of the testes (which makes up almost all of the testicular mass) continued to go through circannual cycles, whereas the testosterone-producing Leydig cells remained continuously active. Since the tubular tissue is predominantly under the control of FSH, the Leydig cells under that of LH, these data suggest that in this particular bird the rhythm in FSH continues, whereas the rhythm of LH ceases. The reverse may have been the case in the bird shown in the lower panel, which had a pronounced circannual rhythm in bill coloration and molt, while the rhythm in testis size was hardly discernible.

5.2.2 Internal Dissociation of Various Circannual Functions

Even stronger evidence for independence among some circannual functions comes from observations indicating that different rhythmic processes may drastically change their internal phase-relationship with each other (e.g., Gwinner 1968a; Berthold et al. 1972a, b; Gwinner and Dorka 1976). In the synchronized state different circannual functions within an individual can alter their phase angle to a changing zeitgeber period at different rates. The result of this is that the temporal relationship between circannual functions is different in different zeitgeber periods (Aschoff 1980). Similar phenomena may also occur spontaneously in constant conditions. Examples of this nature from the circannual cycles of testicular size and molt of garden warblers are shown in Fig. 5.8. The data in the upper diagram (A) are from a typical bird in which each testicular cycle was preceded by a prenuptial molt and followed by a postnuptial molt. The same was true during the first year of the birds B, C, and D in Fig. 5.8. Later in the

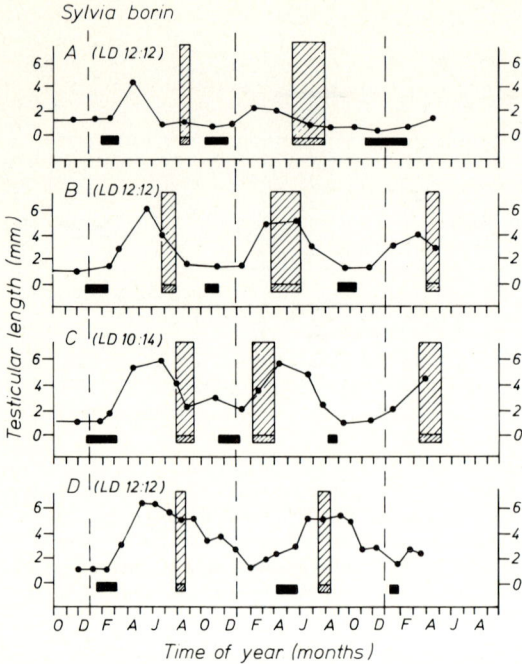

Fig. 5.8. Variations in testicular length and occurrence of molt in four garden warblers (*Sylvia borin*) kept for about 2.5 years under a constant LD 12:12 or LD 10:14. *Solid bars* prenuptial molt; *hatched bars with hatched columns on top* postnuptial molt. (After Gwinner and Dorka 1976)

course of the experiment, however, this normal sequence of events was disrupted. This is most obvious if one compares the testicular cycle with that of postnuptial molt during the second and third experimental year. In all three birds postnuptial molt was advanced relative to the testes cycle, so that it occurred simultaneously with maximum testicular activity (birds B and D) or even during testicular growth (bird C). Hence, large testicular size and molt are not necessarily mutually exclusive, as might have been concluded on the basis of the normal temporal relationship between these two processes.

Even more pronounced changes can occur in the phase relationship between the annual cycles of molt, migratory restlessness and body weight. Results from an experiment on garden warblers are shown in Fig. 5.9. These birds were kept for 33 months in LD 12:12. During the birds' first subjective spring, the temporal relationship between the three functions were normal: the increase in body weight was accompanied by the onset of vernal zugunruhe, both activities being preceded by a complete prenuptial molt. In the subsequent course of the experiment, this relationship changed, so that the birds' third subjective vernal molt coincided with fattening and the onset of zugunruhe did not begin before body weight started to decrease. This indicates that the normal sequence in which these three processes occur is not a result of their being sequentially linked.

It is appropriate here to discuss briefly a possible biological significance of these phenomena. In several species of free-living birds the temporal sequence of annual events is not necessarily rigidly fixed but may be subject to considerable individual or year-to-year variations. This applies to the cycles of breeding and molt in birds inhabiting highly unpredictable environments such as desert (Ser-

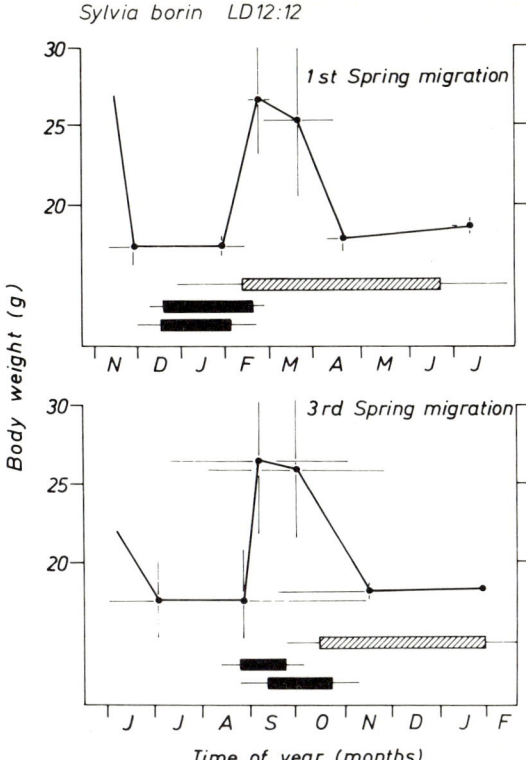

Fig. 5.9. Changes in body weight (*curves*) as well as occurrence of zugunruhe (*hatched bars*) and molt (*upper solid bars* molt of body feathers; *lower solid bars* molt of flight and tail feathers) in a group of garden warblers (*Sylvia borin*) held for 33 months in a constant LD 12:12. *Horizontal lines at the bars* and *horizontal* and *vertical lines at the curve points* represent standard deviations. The *upper panel* shows the situation during the first vernal migration season, the *lower panel* the situation 1.5 years later during the third vernal migration season. The figure is partly identical with Fig. 3.8. (After Berthold et al. 1972a)

venty and Marshall 1957; Farner and Serventy 1960; Immelmann 1963a, b; Keast 1968) or the humid tropics (Snow and Snow 1964; Fogden 1972). "The time of molt here may be quite independent of the time of breeding, sometimes coinciding, sometimes overlapping and sometimes alternating rather than a regular seasonal overlap" (Payne 1972). In some of these species molt is timed to a particular season, whereas breeding may occur at any time of the year, depending on the occurrence of propitious environmental conditions. In other species breeding is seasonal but molt is not. A most compelling example is the long-tailed hermit (*Phaethornis superciliosus*), a tropical humming bird (Fig. 5.10). Birds of this species began to molt almost exactly 1 year after hatching and subsequently at about 12-month intervals. Since the hatching dates in the population varied over 6 months, the molting seasons of different individuals also extended over a period of about half a year. However, all individuals came into breeding condition at the same time of year. As a consequence, there was a large variation in the temporal relationship between the timing of molt and reproductive activity on an individual level: depending on the individual's hatching date, molt did or did not coincide with breeding (Stiles and Wolf 1974). It is not known whether a circannual rhythmicity is involved in the control of molt and reproduction in the long-tailed hermit or in any of the other species mentioned. Nevertheless, the results described above for warblers suggest that the large individual and year-to-year vari-

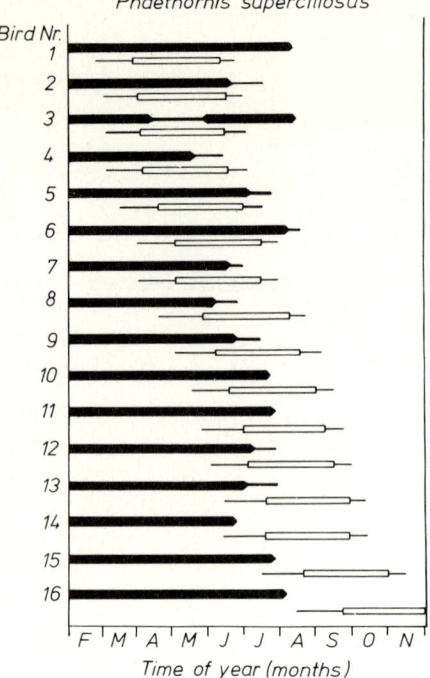

Fig. 5.10. Timing of molt and reproductive activity in 16 males of the long-tailed hermit (*Phaethornis superciliosus*) studied at a lek in Costa Rica. In this species, males defend mating territories and hold singing assemblies during the reproductive season: females come to the assembly to mate. ■ full territoriality; — reduced territoriality; □ heavy molt; —— molt. All individuals maintained mating territories during about the same time of year, but they molted at different times of the year, depending in the date of hatching. (After Stiles and Wolf 1974)

ation in the time relationship between the molt and the breeding cycle of these tropical birds could be the result of changes in the internal phase relationship among different circannual oscillations which are affected differently by environmental conditions. Such variations may also be the basis for photoperiodically induced shifts in the time-relationship between various functions of the kind described in Chap. 6.2.2.6 and for population and species differences in the timing of various functions relative to each other (Chap. 6.2.2.2).

5.2.3 Selective Manipulation of Rhythmic Functions

A final line of evidence in support of the hypothesis that different overt circannual functions may be physiologically independent from each other comes from studies in which one seasonal activity could be experimentally modified without affecting other functions. Such manipulations have been performed at different levels of organization. An example of a manipulation at a peripheral level is shown in Fig. 5.11. Golden-mantled ground squirrels were prevented from depositing fat in autumn. This had no effect on the occurrence and timing of hibernation, indicating a basic independence of the hibernation cycle from the fattening cycle (Pengelley 1968; Heller and Poulson 1970). Conversely, preventing individuals of the same species from hibernation by exposing them to high ambient temperatures had only minor effects on the period of the circannual rhythms of body weight and food consumption (e.g., Pengelley and Fisher 1963; Pengelley

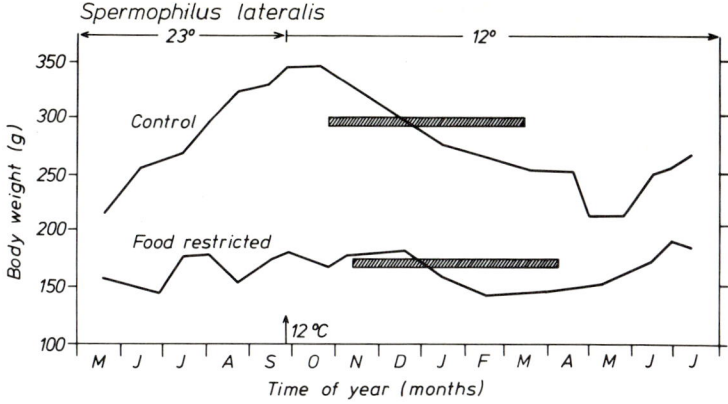

Fig. 5.11. Body weights (*curves*) and hibernation periods (*hatched bars*) of two golden-mantled ground squirrels (*Spermophilus lateralis*) held in a constant LD 12:12 and a temperature of 23 °C (May through September) initially and then 12 °C (September through July). The control animal was allowed to feed ad libitum, whereas the experimental animal received only enough food to keep its body weight at a low (essentially fat-free) level. (After Pengelley 1968)

et al. 1978; Mrosovsky 1978; see also Funakoshi and Uchida 1982 for the bat *Vespertilio superans*). In golden-mantled ground squirrels, food deprivation in winter led to an extension of the hibernation period, but only marginally affected the onset of hibernation in the following year (Fig. 3.2; Mrosovsky 1980b). In garden warblers, the onset and end of autumnal zugunruhe as well as the following onset of winter molt was not influenced by suppressing the body weight increase that normally accompanies zugunruhe. In addition, suppression of autumnal zugunruhe for 2 months by exposing birds to complete darkness at night, did not affect the timing of winter molt (see Gwinner 1977b for a review).

At the level of endocrine regulation there are examples showing that the elimination or modification of hormones affecting one circannual rhythm may have no effect on other circannual rhythmicities. Thus castration of male and female golden-mantled ground squirrels held in LD 12:12 (Pengelley and Asmundson 1969) and of female Turkish hamsters (*Mesocricetus brandti*) held in LD 10:14 (Hall and Goldmann 1982) did not affect the hibernation rhythm. Similarly, the circannual body weight rhythm of golden-mantled ground squirrels held in LD 14:10 was only marginally affected by castration and/or treatment with testosterone or estradiol (Zucker and Boshes 1982). Finally, the circannual rhythm of molt persisted in starlings held in LD 12:12 after castration (E. Gwinner and J. Dittami, unpublished).

Proceeding finally to the level of the central nervous system, results analogous to those obtained at the endocrine level are not yet available. However, lesion experiments carried out by Zucker et al. (1983) with male golden-mantled ground squirrels are consistent with the hypothesis that ablation of the SCN or of adjacent brain areas may result in a partial dissociation between circannual rhythms of reproduction and body weight. In 11 lesioned males maintained for 21 months in continuous light, the average period of the reproductive cycles (measured between the first two onsets of scrotal pigmentation) was 317 days, whereas the average period of the body weight cycles (measured as the interval between the first two body weight troughs) of the same animals was 354 days. Corresponding values for three intact males were 367 days and 350 days respectively. For some lesioned animals the differences between the two periods were large, e.g., 398 vs. 293 days or 356 vs. 258 days, respectively. Zucker et al. (1983) concluded from these results that "hypothalamic damage, in concert with constant illumination interferes with internal synchronization of circannual rhythms; the results suggest that different oscillators underly the circannual rhythms in body weight and reproduction."

5.2.4 Conclusions

The results presented above indicate that different overt circannual rhythms within an organism may be controlled by independent physiological systems, some possibly functioning as separate circannual clocks. In an attempt to make this point, emphasis has been placed on cases that speak in favor of independent systems but, of course, there are many other instances that rather suggest that different rhythms are tightly locked and possibly interrelated. Obviously, the temporal organization in various species can be very different and even within one species different functions could vary in their dependency on one another. Still, it seems necessary to look separately at individual circannual rhythms. The following sections deal with some of the attempts that have been made to identify the components of specific circannual functions.

5.3 Components of Specific Circannual Functions

5.3.1 Attempts at Changing Circannual Period by Altering the Duration of Potential Components of the Cycle

A possible way of identifying the components of a circannual rhythm would be to alter the duration of a particular process suspected to be part of the rhythmicity. For example, one might attempt to change the period of a circannual reproductive rhythm by modifying that fraction of the cycle during which the gonads are active. If the processes responsible for active gonads constitute integrate parts of the whole cycle, this alteration should change the overall period of the circannual rhythm. In line with this, Mrosovsky (1978) pointed out that the reduction of reproductive activities in captive animals may result in a shortening of their sexually active phase, which, in turn, may be responsible for the fact that the period of many circannual reproductive rhythms is shorter than one year (Tables 2.1–2.3). This idea is supported by the finding in golden-mantled ground squirrels that the circannual period measured between two successive hibernation phases was considerably longer in females that went through gestation and lactation in spring than in females that did not. During the following cycles, when none of the animals became pregnant, the average circannual periods of the two groups were indistinguishable but the phase difference between the rhythms of the two groups, caused by the lengthening of the first period in the group with litters, was maintained (Pengelley and Asmundson 1975). Other, more circumstancial evidence supporting this idea has been summarized by Mrosovsky (1970).

5.3.2 Attempts at Identifying External and Internal Conditions Under Which Rhythmicity Stops or Continues

Another promising approach toward identification of the nature of the stages that compose circannual cycles makes use of the fact that the conditions under which circannual rhythms are expressed are often very limited and different in different species (Chap. 3.1). The identification of the processes that continue under one set of conditions, and, hence, sustain the rhythm, but not in others, and, hence, presumably stop the clock, may help to identify the physiological nature of particular components of the cycle.

This possibility has been exploited in experiments carried out with the European starling. As schematically shown in Fig. 5.12 the circannual testicular cycle of this species continues only in LD 12:12. In this condition, the birds go through all the stages characteristic for an annual cycle of gonadal function (Gwinner et

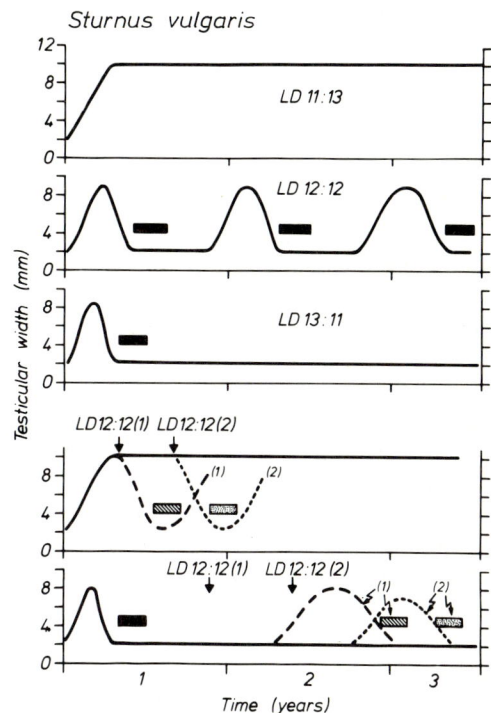

Fig. 5.12. Upper graph A schematic representation of gonadal growth and regression *(solid lines)* and molt *(bars)* in male European starlings *(Sturnus vulgaris)* held in three different photoperiodic regimes: LD 11:13, 12:12, and 13:11. *Lower graph* The effect of placing starlings from either an LD 11:13 or 13:11 in an LD 12:12. (Gwinner and Dittami 1986)

al. 1985b). In LD 11:13 or any shorter photoperiod the gonads grow but then remain large and active for years. In LD 13:11 or any LD cycle with longer photoperiod the gonads go through an initial cycle, but then stay small and inactive indefinitely.

This situation raises the question of whether the clock system controlling the gonadal rhythm is indeed arrested under long and short photoperiods. A priori, the possibility cannot be excluded that a circannual clock keeps running under these conditions, but that its overt expression is obscured as a result of the prevailing photoperiodic conditions or "masked", to use a term introduced by Aschoff (1960) to describe situations in the circadian field in which overt rhythms are not discernible although the underlying circadian oscillation continues. However, the results summarized schematically in the lower part of Fig. 5.12 exclude such a possibility. In these experiments starlings whose testicular cycles were arrested under an LD 11:13 or under an LD 13:11 were moved after variable times from these conditions to an LD 12:12 in which rhythmicity was resumed. The time that elapsed between the transitions to LD 12:12 and the occurrence of subsequent events turned out to be basically independent of the time span of long days to which the birds had previously been exposed. This would not have been expected if an underlying clock had kept running in LD 11:13 and LD 13:11. In this case the transitions would have occurred at different circannual phases and differential time courses would have been expected for the processes following the transfers.

Subsequent experiments have revealed that the arresting of the rhythm under short photoperiods is due to the failure of the birds to enter the state of photorefractoriness, which requires a long photoperiod (Falk and Gwinner 1983). Conversely, the arresting of the rhythm under long photoperiods was shown to be due to the failure of the birds to recover from photorefractoriness, which requires a short photoperiod (Gwinner and Wozniak 1982). In a 12-h photoperiod both of these transitions are possible and correspondingly both long- and short-day birds resume cyclicity after transfer to LD 12:12 (Fig. 5.12).

The termination and the initiation of refractoriness are then crucial events in the chain of processes that make up a circannual gonadal cycle in the starling. The next step in the analysis of this rhythm must therefore consist in the identification of the physiological processes that are involved in the termination and induction of refractoriness. Almost nothing is known as yet about the processes that *terminate* refractoriness, although all available evidence suggests that the important mechanisms are localized at a relatively high level of the hypothalamo-hypophysial-gonadal axis or higher, and are, for instance, largely independent of gonadal hormones (e.g., Matt 1982; Farner et al. 1983; Goldsmith and Nicholls 1984b).

The *induction* of refractoriness is probably also achieved by mechanisms that occur at a relatively high level of the control system, although there is now a considerable amount of data indicating that thyroid hormones may be crucially involved. This is suggested by the finding that thyroidectomy early during the testicular growth phase prevents the development of refractoriness in a long photoperiod (Wieselthier and van Tienhoven 1972; Goldsmith and Nicholls 1984a), but subsequent treatment with thyroxine induces it (Goldsmith and Nicholls 1984a;

Nicholls et al. 1985). The latter effect is not the result of an inhibition at the level of the gonads, since treatment with thyroxine also inhibits gonadotrophin secretion in castrated starlings held in LD 11:13 (Goldsmith and Nicholls 1984b). Thyroxine may exert its effects through stimulating prolactin secretion, as suggested by the following results:

1. In wild starlings plasma prolactin levels are maximal at about the time at which the birds become refractory (Dawson and Goldsmith 1982). In captive starlings, held in two long photoperiods that induced refractoriness at different rates, maximum prolactin levels also occurred just before the onset of refractoriness (Dawson and Goldsmith 1983).

2. In contrast to control birds, thyroidectomized starlings, which under long photoperiods do not become refractory, also show no increase in prolactin, as do the controls (Goldsmith and Nicholls 1984c).

3. Prolactin levels increase, however, if such thyroidectomized birds are treated with exogenous thyroxine, a procedure that induces refractoriness (Goldsmith and Nicholls 1984a). Thus, prolactin secretion appears to be "always initiated under circumstances which would eventually lead to photorefractoriness. Is the hormone directly involved in causing refractoriness? We do not yet know." (Nicholls et al. 1985).

This section has concentrated on the annual reproductive cycle of the European starling as a model system for the approach that aims at understanding the physiological nature of circannual rhythms by defining the external and internal conditions under which rhythmicity can be arrested and re-initiated. It is obvious that the same kind of approach should also be feasible for analyzing other cases listed in Tables 2.1–2.3, in which the range of environmental conditions permissive for the expression of circannual cycles is limited. However, although a lot is known about the environmental and physiological constraints of particular phases of the annual cycles in some of these animals (e.g., hamsters, sheep) only little effort has yet been made to use such information for the analysis of circannual systems as a whole.

5.3.3 Identification of Hormones and Central Nervous Structures Involved in the Control of Circannual Rhythms

A third way of analyzing circannual rhythms and their components obviously consists in the direct identification of the endocrine glands and/or the central nervous structures involved, by the classical techniques of elimination or substitution. These methods allow both the identification of the level(s) at which the oscillatory processes occur and the determination of the physiological components that play a role. A few investigations of this kind have been carried out on avian and mammalian circannual reproductive cycles.

a) Gonads

There is evidence for both birds and mammals indicating that the gonads and their hormones are not integrate components of the basic circannual reproductive rhythm. In castrated male starlings held in LD 12:12, a circannual rhythm in plasma LH continued with an amplitude larger than, but a period similar to, that of intact controls (E. Gwinner and J. Dittami, unpublished). Female ovariectomized golden-mantled ground squirrels held in constant LD 14:10 also exhibited circannual cycles in LH, the periods of which were close to the circannual periods of intact conspecifics (Zucker and Licht 1983a). Castrated male ground squirrels held in the same conditions, in contrast, had chronically elevated LH levels with no clear-cut rhythms (Zucker and Licht 1983b). These results suggest that in female ground squirrels, as in male starlings, the onset and end of gonadotrophin secretion by the hypothalamo-hypophysial system is controlled by a circannual rhythmicity that functions independently of gonadal secretions, whereas in male ground squirrels the endogenous rhythmicity works by altering the sensitivity of the hypothalamo-hypophysial axis to the negative feedback effects of gonadal hormones. Sex differences in the responsiveness to gonadectomy are also suggested by results obtained from sheep in which, opposite to the situation in squirrels, plasma LH levels remained elevated in gonadectomized females but went through an annual cycle in males (e.g., Lincoln and Short 1980; Legan and Winans 1981). The investigations on sheep, however, have been performed under changing photoperiods, so that conclusions about the situation in constant conditions are not yet possible.

b) Pineal Gland

Birds. Although most data indicate that the pineal plays only an insignificant role in the control of annual cycles in most birds (Gwinner et al. 1981), pinealectomy had some effects on the expression of the circannual testicular cycles in starlings held in constant conditions: in LD 12:12, the proportion of male birds initiating a second testicular cycle was significantly lower in pinealectomized birds than in controls. In constant dim light, in contrast, the pinealectomized birds showed a clearer circannual testicular cycle than the controls (Fig. 5.3). Whether pinealectomy in these birds was effective through changing the circadian photoperiodic time-measuring system or in some other way cannot be evaluated at present (see p. 74 for a discussion).

Mammals. The central role of the pineal organ as a photoneuroendocrine transducer has been well established for a variety of mammalian species (Reiter 1974, 1980; Goldman and Darrow 1983 for reviews), but its participation in the control of circannual rhythms under constant conditions has hardly been studied as yet. In both ferrets (*Mustela putorius*: Herbert 1971, 1972) and golden-mantled ground squirrels (Zucker 1985), pinealectomy did not abolish circannual cycles in constant conditions: ferrets held for 2 years in an LD 14:10 went through two successive estrous cycles about 17 months apart. This rhythm was expressed even more clearly than in sham-operated animals, but tended to have a slightly longer period. Similarly, in golden-mantled ground squirrels a circannual rhythm in body weight and reproductive condition persisted in pinealectomized animals.

The period was slightly shorter than that of the controls. The annual rhythms of body weight and hibernation in pinealectomized golden-mantled ground squirrels held under simulated natural photoperiodic variations were also only marginally different from those of control animals (Phillips and Harlow 1982; Ralph et al. 1982).

Although pinealectomy in ferrets did not abolish the circannual reproductive rhythm of animals held under constant conditions, the same operation drastically altered its pattern under natural lighting conditions. In the control and sham-operated animals estrous occurred at about the same date in two consecutive years. The second estrous of pinealectomized animals, however, was delayed by about 5 months (Herbert 1971, 1972; Herbert et al. 1978). Hence the period of the reproductive cycle of these animals was similar to that observed in the animals held in constant conditions. These results, then, are consistent with the hypothesis that pinealectomy has only minor effects on the circannual mechanism as such but can prevent its synchronization by the annual photoperiodic cycle. Similar conclusions have been drawn from results on sheep indicating that the annual reproductive cycle of pinealectomized ewes could (in contrast to controls) not be synchronized with a 180-day rectangular photoperiodic cycle, but rather showed a period close to one year under these conditions (Bittmann et al. 1983a; see also e.g., Lincoln 1979; Bittmann et al. 1983b for long-term pineal effects in sheep). Indeed, the behavior of these pinealectomized animals was similar to that of blinded ewes in similar conditions (Legan and Karsch 1983).

c) Central Nervous System

Up to now no neural structure has been identified that is essential for the generation of circannual rhythms. In golden-mantled ground squirrels, circannual rhythms of body weight, food consumption, and hibernation persisted with an unchanged period after lesioning the ventromedial hypothalamic area. These lesions only produced a marked increase in the level and amplitude of the body weight cycle and, in some cases, slight phase shifts of the rhythms (Mrosovsky 1975). Similar results were also obtained in thirteen-lined ground squirrels (Mrosovsky 1974a). Ablation of the paraventricular nucleus of golden-mantled ground squirrels did not affect the circannual rhythms of body weight and reproduction (Dark and Zucker 1986). Lesions applied to the midbrain central gray region of the same species reduced the amplitude of the body-weight cycle, but left the rhythm otherwise intact. Ablations of the suprachiasmatic nucleus (SCN) in golden-mantled ground squirrels had no effect on the circannual body-weight rhythm in 7 out of 11 animals. Only in two animals was the rhythmicity eliminated, in one it was phase-shifted, and in yet another the circannual body-weight rhythm was replaced by 3- to 5-months cycles (Zucker et al. 1983). These and other similar data obtained from the same species (Dark et al. 1985) indicate that intact SCNs are not necessarily required for the persistence of the circannual body-weight cycle, but that rhythmicity may be affected under certain conditions by SCN lesions. In addition, SCN lesions apparently affected the normal coupling between the circannual rhythms of body weight and reproduction (see Sect. 5.2.3). In European hamsters (*Cricetus cricetus*), finally, lesions of the ascending noradrenergetic bundles and of the central gray lateral to the dorsal raphe nuclei had slight effects on the amplitude and phase of the body-weight rhythm, but the basic rhythmicity was maintained with the normal period (Canguilhem et al. 1977).

5.4 Conclusions

Although circannual rhythms have been known for more than two decades, only little has been learned as yet about their physiological basis. Indeed, it appears from the investigations and considerations summarized in the preceding sections that the field is still struggling with the problem of defining questions that are specific to it. It may be appropriate, therefore, to elaborate briefly at the end of this chapter on the problem along which lines future research on circannual mechanisms might be directed.

On a formal level, it is still not clear how – if at all – circadian rhythms interact with the circannual system. Since circadian clocks are used by many organisms for photoperiodic time measurement, it might be expected that the circadian system mediates circannual entrainment in species that use the annual photoperiodic cycle as a circannual zeitgeber. Good evidence for this prediction is still lacking. However, results obtained from ferrets and sheep indicating that pinealectomy affects circannual entrainment without apparently abolishing circannual rhythmicity are interesting in this context (Sect. 5.3.3). In many mammals the pineal is intimately involved in transducing photoperiodic information to hypothalamic control systems; hence, pinealectomy presumably disconnects the photoperiodic clock from the circannual mechanism. Future studies might approach this problem more directly by asking whether the abolition of circadian pacemakers like the SCN in mammals, the pineal in some birds, or the lobi optici in insects interferes with circannual photoperiodic entrainment. Results on golden hamsters have already shown that SCN-lesioned animals can no longer measure photoperiod appropriately at some stages of their annual cycle (Rusak and Zucker 1979; Turek 1983; Hastings et al. 1985), but whether the same is also true for animals with circannual rhythms remains to be demonstrated.

Apart from its possible role in circannual synchronization, the circadian system might be involved in the generation of circannual rhythms. Several lines of evidence summarized in Sect. 5.1 indicate that those few circannual rhythms that have been studied so far do not depend in a simple way on the circadian system. However, it is still possible that there are more subtle effects, which in view of our limited knowledge about both classes of rhythms, cannot as yet be easily evaluated and formulated as testable hypotheses. Certainly, the question raised by the work of Meier and his co-workers (Sect. 5.1.3), whether changes in the phase relationship between circadian rhythms in neuroendocrine function may be the basis of circannual changes, should be investigated further.

Apart from the problem of circadian-circannual interactions, two sets of questions should, in my opinion, be central to the future study of circannual mechanisms. The first stems from the possibility raised by Mrosovsky and others that circannual rhythms did perhaps "not evolve as a coherent whole but rather as a series of disjunct parts that became fused when the transitions between states became spontaneous." (Joy and Mrosovsky 1985). If this were true, an appropriate question would be: how is this fusing of states achieved? It can be approached by trying to define the environmental conditions in which the transitions from one state to the next does or does not occur, and by looking simultaneously at the physiological changes which are associated with that transition or its blockage. Alternatively, comparative studies of related taxa which in one particular condition do or do not show the transition could be helpful. In the European starling, the former kind of approach has led to the formulation of concrete questions. One derives from the observation that in this species, depending on circannual phase, a 12-h photoperiod is interpreted as being either just a long or just a short photoperiod; it is therefore obviously central to the understanding of this circannual rhythmicity to find out which internal processes are responsible for these spontaneous changes in interpretation of daylength (cf. Robinson and Follett 1982).

Perhaps the most formidable physiological problem in circannual rhythm research arises from the extremely long duration of the processes involved. Some of the changes that occur within or between the various fractions of a circannual cycle have time constants that are way beyond the range of time constants known for any neuroendocrine feedback loop and are in many respects reminiscent of developmental processes occurring during ontogeny, e.g., sexual maturation. What is the nature of the changes that take place in a migratory warbler that spends the winter in a constant equatorial photoperiod to which it does not respond with gonadal growth in November but to which it eventually does respond in late March? And which processes are responsible for the slow and spontaneous variations in the metabolic set-point that determines the body weight cycle of a hibernating ground squirrel in a seasonally constant environment? The scanty evidence summarized in Sect. 5.3.3 suggests that these changes occur at a relatively high level of the system, presumably in the brain. It seems possible that they involve neuronal growth and degeneration processes of the kind recently shown to occur in connection with seasonal song learning of canaries (Nottebohm 1981).

Chapter 6

Adaptive Significance of Circannual Rhythms

6.1 General Advantages of Circannual Rhythms

The advantages of circannual control systems, as opposed to systems that depend exclusively on external proximate factors are by no means obvious. Circannual rhythms function only in connection with external seasonal zeitgebers that synchronize them with the natural year and hence the question arises whether these external factors could not do the same job alone. As with circadian rhythms (Enright 1970) there appears to be no convincing answer to this question as yet, but two general hypotheses about the biological significance of circannual rhythms merit special consideration. They will be discussed in the following sections (6.1.1 and 6.1.2).

6.1.1 Improvement of Consistency of Seasonal Timing

Although experimental evidence is missing, it can be assumed that a circannual rhythmicity acts as a buffer system between the environment and the physiology of an organism, that protects seasonal functions in organisms against environmental noise. In contrast to passive systems or to highly damped oscillators, self-sustaining oscillators are "inert", so that day-to-day variations in external conditions will be averaged out by the system, resulting in an increased consistency of seasonal activities between years. This may be of particular significance for organisms that inhabit environments with pronounced annual changes where the time spans available for the various seasonal activities are short. If the organisms relied only directly on external cues, a considerable year-to-year variability in the timing of their seasonal activities would be expected as a result of the variability of weather conditions, with the result that they might miss the optimal season. Even photoperiod, the most reliable of the known annual timing cues, is not free from variations, and the seasonal information provided by it can be considerably falsified, for instance by overcast skies, which at higher latitudes can reduce the effective photoperiod by several hours (Klein 1972). If this were true, circannual mechanisms would be particularly necessary for organisms that are most dependent on the precise timing of their seasonal activities (see Gwinner and Dittami 1986, for a discussion). As will be shown in the following sections, this prediction is indeed supported by the results of comparative investigations on closely related species (Gwinner 1981d).

6.1.1.1 Hibernating Mammals

In the obligatorily hibernating golden-mantled ground squirrel (*Spermophilus lateralis*), that inhabits the boreal life zone of North America, the annual rhythms of body weight and hibernation are most rigidly endogenously controlled. Under constant conditions of photoperiod and temperature, these rhythms persist undamped for at least five cycles. The closely related Mohave ground squirrel (*S. mohavensis*) and the round-tailed ground squirrel (*S. tereticaudus*), that inhabit arid environments, depend to a lesser degree on such an endogenous rhythm. In constant conditions, their body weight and hibernation rhythms show tendencies to dampen (*S. mohavensis*) or even disappear during the first cycle (*S. tereticaudus*). Pengelley and Kelly (1966) emphasized that a circannual rhythm is an excellent adaptation for *S. lateralis* that inhabits a severe and regularly changing environment "where each event in the animal's life cycle must occur at an exact time in order for it to survive." Particularly, a circannual rhythm guarantees that emergence from the hibernation burrow occurs at the proper time, i.e., early enough for successful reproduction and late enough to avoid unfavorable conditions in late winter and early spring. Heller and Poulson (1970) pointed out that for high-latitude species "it would not be adaptive to have a flexible breeding date. An early spring warm spell might be followed by severe cold; and on the other hand, it would not be advantageous to delay breeding because of a late spring if the time remaining for the young to mature and prepare for winter were insufficient. In such an environment it would be important to breed on the average optimal date regardless of meteorological conditions, and therefore, selection would favor precision rather than flexibility." An endogenous rhythmicity, due to its inherent inertia, might be an ideal device for the necessary averaging. For other species of *Spermophilus*, particularly such that inhabit arid environments like *S. mohavensis* and *S. tereticaudus,* flexibility may be at a premium partly because in these areas environmental conditions are less predictable from year to year and partly because optimal conditions, once they occur, are not necessarily as restricted as for the higher-latitude species. Hence, the chances for successful reproduction are probably higher in individuals that can adjust their behavior to prevailing conditions as compared with the strategy of breeding at an average optimal date. Consequently rigid circannual rhythms have not evolved in these species.

A relationship between the predictability of climatic conditions and the degree of involvement of a circannual rhythmicity has also been demonstrated for chipmunks. Heller and Poulson (1970) found that among four related species inhabiting different altitudinal zones of the eastern slope of the Sierra Nevada in California, the two species (*Eutamias alpinus* and *E. speciosus*) from the upper zones, with their rigorous climatic variations, had more clearly manifested endogenous rhythms in the laboratory than the two species (*E. amoenus, E. minimus*) from the lower zones, with their less rigorous and more variable climates.

Less rigid control by a circannual rhythmicity may be related to the particular food habits of an animal. In California, the belding ground squirrel (*S. beldingi*) occurs sympatrically with *S. lateralis* and experiences the same climatic conditions. *S. beldingi* is specialized on succulent vegetation throughout the year.

S. lateralis only feeds on succulent plants in spring and summer, but in fall it turns largely to seeds to build up fat depots. It also stores seeds in the burrow as an extra food reserve for the winter. Food encachment is not possible for *S. beldingi,* because the succulent plants on which it depends entirely are not suitable for storage. Hence, *S. beldingi* must rely exclusively on fat storage, and since the availability of succulent plants varies from week to week "*S. beldingi* must avail itself of food whenever possible and conserve its fat when food is not available". Therefore, "...it is not adaptive for it to have as strong an endogenous control of its annual cycles as does *S. lateralis.*" Consistent with this, the circannual rhythm of *S. beldingi* is less rigid and more susceptible to environmental changes than that of *S. lateralis* (Heller and Poulson 1970).

6.1.1.2 Migratory Birds

Comparative studies of European migratory birds have revealed a rigid and persistent circannual control of annual functions in long-distance migrants wintering at or close to the equator. In closely related species that are short-distance migrants, wintering in the Mediterranean area or in northern Africa, less pronounced circannual rhythms have been found. A good example is provided by the closely related species willow warbler (*Phylloscopus trochilus*) and chiffchaff (*Ph. collybita*), which breed sympatrically over a large range of central and northern Europe, but have widely different wintering areas. Willow warblers that breed in Germany, winter at or beyond the equator, whereas German chiffchaffs migrate only to the Mediterranean countries and northern Africa (Fig. 6.1). If held under a constant 12-h photoperiod and at constant temperature conditions a pronounced circannual rhythmicity in molt, zugunruhe and body weight was found in the willow warblers with patterns similar to those held under a naturally varying photoperiod (Fig. 6.2). This rhythm in fact continued over an experimental period of 27 months (Fig. 2.7). In chiffchaffs held under the same conditions, in contrast, a relatively normal seasonal pattern was observed only during the first year, but then rhythmicity became irregular or disappeared (Fig. 6.2, and Gwinner 1971a).

During the first year, when both species maintained a normal pattern, interindividual variability with regard to both timing and amplitude of seasonal activities was consistently higher in the chiffchaffs than in the willow warblers. As shown in Fig. 6.3, changes in body weight during the first migratory season varied widely among individual chiffchaffs under both natural and constant photoperiods. Even among siblings, maximal weights differed by as much as 40%. In willow warblers, in contrast, the weight patterns were rather similar and maximal weights of siblings never differed by more than 25%.

Similar differences in interindividual variability were found with regard to almost every parameter studied (Fig. 6.4). The coefficients of variation of duration, amount and maximal value of zugunruhe, both in autumn and spring, were consistently higher in chiffchaffs than in willow warblers. The age at which the second period of plumage development and post-juvenile molt began and ended were more variable in the chiffchaffs than in the willow warblers. The same was true for the duration of these processes. It is proposed hat the higher variability ob-

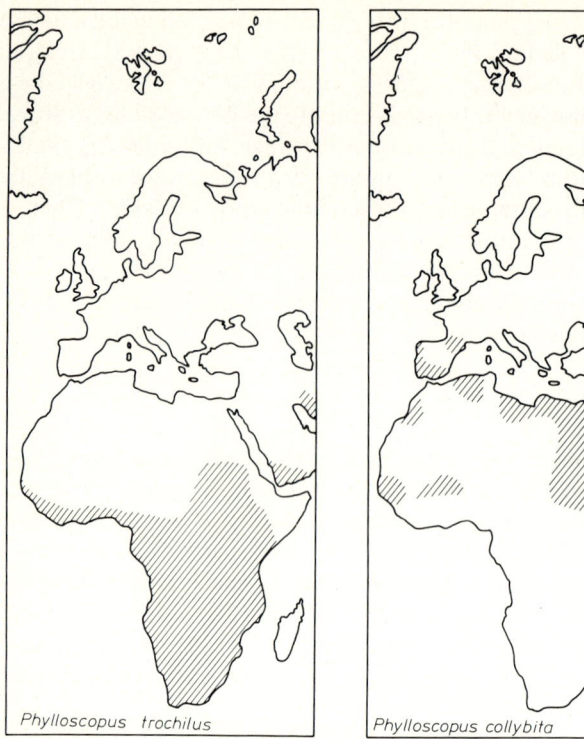

Fig. 6.1. Wintering grounds of the willow warbler (*Phylloscopus trochilus*) and the chiffchaff (*Ph. collybita*). After Dementiew and Gladkow 1954)

served in chiffchaffs results from the fact that the circannual control of both the timing and the amplitude of these processes is less rigid in the chiffchaffs than in the willow warblers.

A similar relationship between rigidity of overt circannual rhythmicity and consistency in the timing of seasonal events among individuals, and the distance travelled by a species or subspecies was observed in comparative studies of sylviinine warblers (Berthold et al. 1972a; Berthold 1974a, b), flycatchers of the genus *Ficedula* (Gwinner and Schwabl-Benzinger 1982) and stonechats (*Saxicola torquata:* Gwinner and Dittami 1985). These observations are consistent with the idea that the existence of pronounced circannual rhythms in long-distance migrants may be related to their much tighter annual time schedule resulting from

---

Fig. 6.2. Variations in zugunruhe, body weight and molt in willow warblers (*Phylloscopus trochilus*) and chiffchaffs (*Ph. collybita*). *Upper two diagrams*: Data from birds held for 19 months under natural photoperiodic variations of their breeding grounds. *Lower diagrams* data from birds that were transferred between September 15 and 20 (*vertical line*) to a constant 12-h photoperiod (LD 12:12). ↓ date of hatching. ▨ zugunruhe: Number of 0.5-h intervals during which a bird was active each night; mean values for successive thirds of a month are plotted. ■ molt of large wing and tail feathers. ▨ molt of body plumage. —— body weight. + animal died. (After Gwinner 1971a)

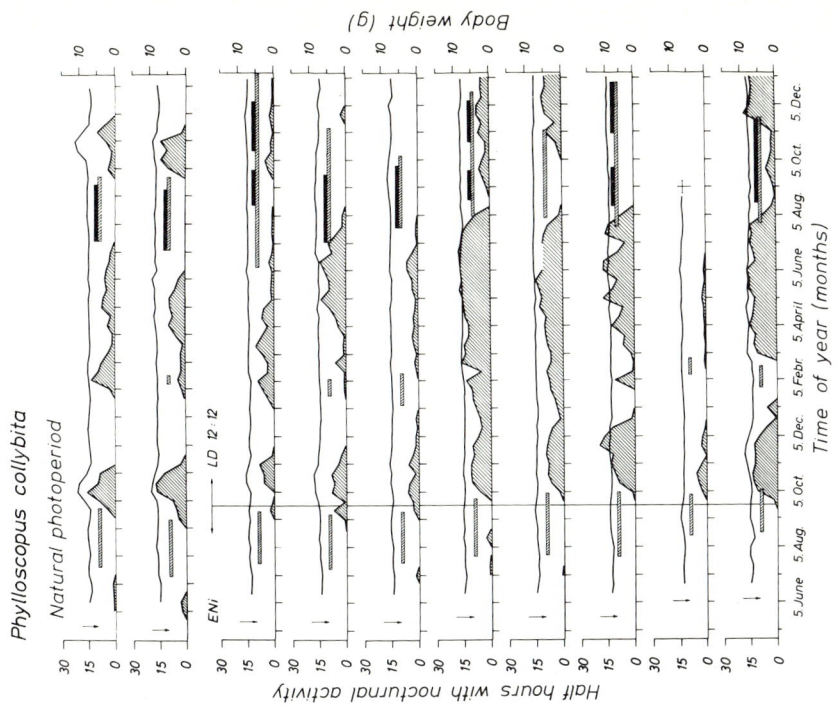
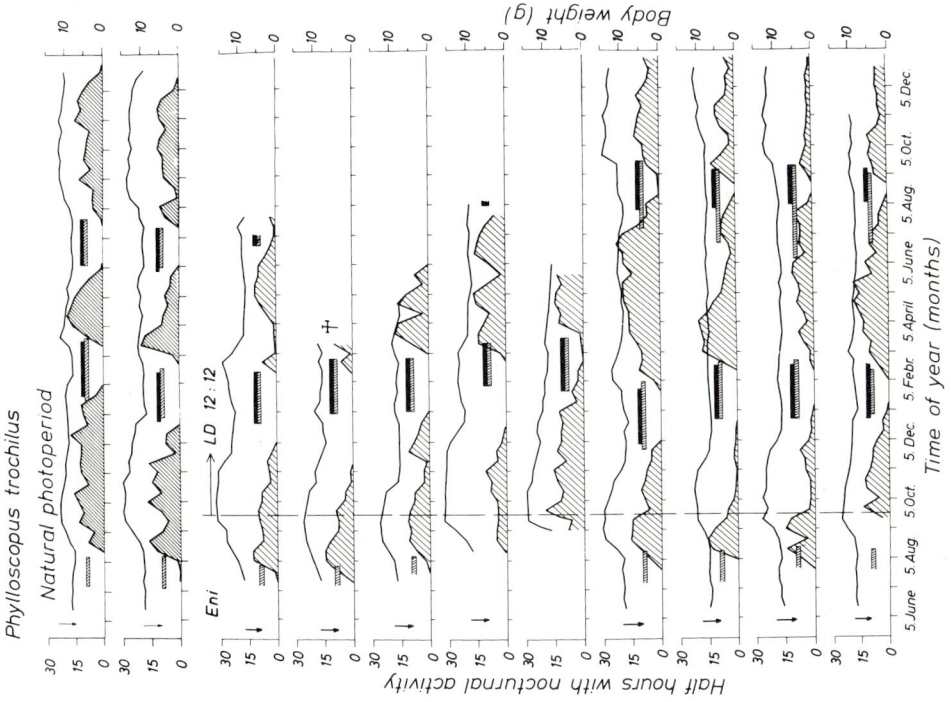

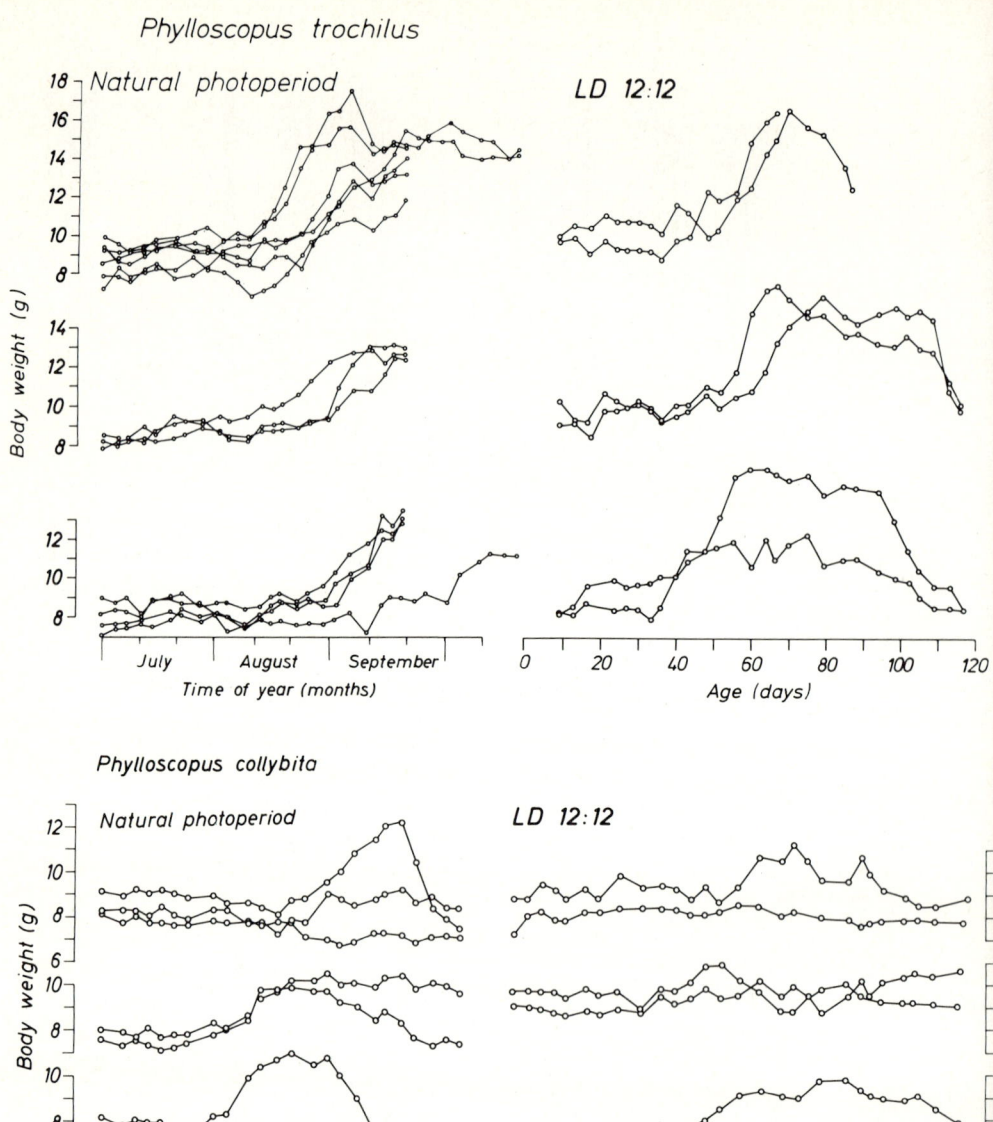

Fig. 6.3. Variations in body weight before and during the autumnal migratory season in first-year willow warblers (*Phylloscopus trochilus;* upper diagram) and chiffchaffs (*Ph. collybita;* lower diagram). Natural photoperiod: Data from birds kept under natural photoperiodic variations of their breeding grounds. LD 12:12: data from birds raised and subsequently kept in a constant 12-h photoperiod. In each set of curves individual weights of siblings are combined. (Gwinner 1972b)

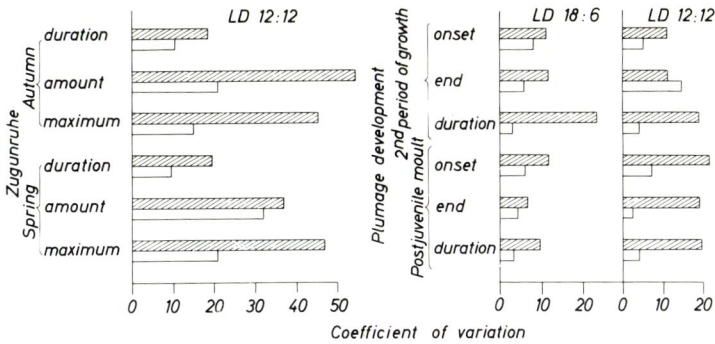

Fig. 6.4. Interindividual variability expressed by coefficients of variation of several parameters of zugunruhe (*left*) and plumage development (right) in chiffchaffs (*Phylloscopus collybita*) and willow warblers (*Ph. trochilus*) kept in a constant 12-h photoperiod (LD 12:12) or 18-h photoperiod (LD 18:6), respectively. ▨ chiffchaffs; ☐ willow warblers (11 > n > 8). (Gwinner 1972b)

the long migrations that require many months each year. As in hibernating mammals in extreme climatic conditions, every event in the annual cycle of these birds must be timed accurately, each year at the same time, for it to be successful. Vernal migration, for instance, must (and indeed does) commence exactly at the right time; that is, early enough for the birds to arrive in their breeding area before territories and mates are occupied by competitors, and before environmental conditions suitable for successful reproduction have deteriorated – and late enough to avoid the unfavorable conditions prevailing in early spring. Hence, these species of long-distance migrants cannot afford to miss the exact time and must therefore be less responsive to day-to-day variations in external conditions in the wintering grounds and along the migratory route. It is suggested that a rigid circannual rhythmicity by virtue of its inertia serves this function of stabilizing seasonal cycles. For the short-distance migrants, in contrast, it may be more adaptive to be flexible as they not only have more time available for their seasonal activities they also migrate later in autumn and earlier in spring than the long-distance migrants and hence are more likely to become confronted with adverse environmental conditions to which they must adjust. The suspicion that short-distance migrants do indeed respond more readily to exogenous factors in the control of their annual cycles is in fact supported by field observations indicating, for instance, that the time of departure in autumn or the times of arrival in spring vary more from year to year in short-distance than in long-distance migrants (Dorka 1966).

6.1.2 Timing of Seasonal Activities in Unpredictable or Constant Environments

The hypothesis developed in the previous section, was based on correlations between the rigidity of the circannual control system of a species and one aspect of its ecology, the suspected need for precise timing. However, there is another correlation that may be more relevant. In mammals, the hibernators with strong circannual rhythms are not only those with the tightest seasonal time schedule, but at the same time those that are the most typical hibernators which spend long periods of time each year in underground burrows under relatively stable ambient conditions. Similarly, the long-distance migrating birds studied so far are all species that winter close to the equator, that is, in areas deficient in reliable seasonal timing cues. Still, these birds must begin migration at the appropriate time each spring. For them a circannual clock is obviously advantageous and does indeed control the onset of migration (see Sect. 6.2.2). It seems possible, therefore, that circannual rhythms have evolved because they provide temporal information for organisms that inhabit seasonally constant or unpredictable environments for a substantial part of the year but must nevertheless time their seasonal activities accurately with respect to pronounced seasonal changes that they encounter at other times of the year.

There are at present no data on the basis of which one could falsify either or both of the two hypotheses presented in this and the previous section. The problem is that in the hibernators, as well as in the migrants, a rigid circannual system is correlated with both factors, tight seasonal time schedule and extended period of time in a relatively constant environment. It should also be kept in mind that the two hypotheses are, of course, not necessarily mutually exclusive. The results and considerations presented above do at least indicate, however, that the degree to which a circannual rhythmicity is involved in the control of seasonal activities may differ among species, even within a genus. Indeed, the ecological conditions that favored the evolution of circannual rhythms may have been very different in various groups of organisms. Hence these considerations suggest that circannual rhythms may be special mechanisms that have evolved not only independently (Chap. 2), but also under different environmental pressures in various groups of organisms.

6.2 Specific Functions of Circannual Timing Mechanisms

In the previous paragraphs I attempted to respond to the general question of the selection pressures that might have been responsible for the evolution of circannual rhythmicities. In the following, attention will be focused on the specific functions effected by circannual rhythms. The question of why, a circannual rhythmicity and no other mechanism, e.g., a system depending directly on external factors or an hour-glass timer, is used to fulfill these tasks will no longer be considered in this context.

6.2.1 Timing and Adaptive Programming of Seasonal Activities in Hibernating Mammals

As outlined in the previous section, circannual rhythms are fundamentally involved in the appropriate timing of hibernation and reproduction in hibernating mammals, and, particularly, in assuring emergence from hibernation at approximately the appropriate time in spring. These functions of circannual rhythms are clearly documented by the examples shown in Figs. 2.1–2.3 and need no further elaboration in the present context.

There is evidence indicating that even details of the seasonal events observed in free-living animals are partly determined by the endogenous time structure of the circannual rhythmicity. For instance, species or population differences in the temporal organization of hibernation and other functions may be reflected in the behavior of animals held under constant laboratory conditions. In a comparative study of five species of ground squirrel exposed to LD 12:12 and either 3 °C or 12 °C, Pengelley and Kelly (1966) found that (as in the field) the "strong" hibernator, *Spermophilus lateralis,* had longer hibernation periods than the "weak" hibernators, *S. mohavensis, S. beecheyi, S. variegatus* and *S. tereticaudus.* The hibernation periods were also less frequently interrupted by active states in *S. lateralis* than in the other less obligatory hibernating species.

Comparing thirteen-lined ground squirrels (*Spermophilus tridecemlineatus*) from an area characterized by a relatively long and harsh winter (Michigan) with conspecifics from an area with a milder winter climate (Kansas), Joy (1984) obtained results suggesting that even population differences in seasonal cycles may result from differences in endogenous circannual programming. When animals from both populations were held in the same constant environmental conditions (Table 2.2), the animals from Michigan reached their first maximum body weight earlier, entered the subsequent weight loss phase more simultaneously and had a relatively longer weight loss phase than the animals from Kansas. This can be interpreted as indicating that the animals from the area with the more severe climate (Michigan) are programmed to enter hibernation earlier and more promptly and to hibernate longer than the animals from the milder area (Kansas). Endogenously programmed population differences in seasonal patterning are also suggested by results obtained in a study on yellow-bellied marmots (*Marmota flaviventris*) from a montane and lowland population (Ward and Armitage 1981).

Even differences between the sexes in temporal behavior are partly preprogrammed by a circannual rhythmicity, as suggested by the results obtained in golden-mantled ground squirrels (Pengelley et al. 1979). They found that males held in a seasonally constant environment arose earlier than females, a difference that exists in free-living populations as well (Hock 1955, 1956; Michener 1984). Males also had shorter hibernation periods than females, but once hibernating they awoke less frequently and hence had longer, continuous periods of hibernation than females. Whether these latter differences are also found in free-living animals remains to be elucidated.

Taken together, all of these results suggest that the timing of seasonal activities in hibernating mammals is partly determined in a species-, population-, and even sex-specific manner by the endogenous circannual system. This is not to say that the temporal sequence of events observed in the field is exclusively endogenously preprogrammed, as environmental factors, of course, do modify the endogenous pattern and often in species-, population-, and even sex-specific ways. The above data do indicate, however, that a significant component in the determination of the specific temporal behavior of these animals is an endogenous circannual clock.

6.2.2 Timing and Adaptive Programming of Seasonal Activities in Migratory Birds

6.2.2.1 Onset of Vernal Migration

As elaborated in the introduction, it was proposed long ago that endogenous mechanisms may be significantly involved in the timing of annual migrations of birds that winter in unpredictable or uniform tropical environments that provide little external information about season. This suspicion had received some early experimental support by the studies of Merkel (1963) and Zimmermann (1966), who showed that spring migratory restlessness in whitethroats (*Sylvia communis*) and dickcissels (*Spiza americana*) developed even in individuals held in a seasonally constant environment. Their results were later confirmed and extended in other studies like that shown in Fig. 6.5. Here the results of an experiment with European willow warblers (*Phylloscopus trochilus*), equatorial migrants, are plotted. Three groups of birds were hand-raised in June, and initially held under a natural photoperiod. In mid-September, the time when most free-living conspecifics have left the European continent on their autumnal migration and are approaching their African wintering areas, the birds were subdivided into three experimental groups: Group 1 (upper diagram) remained under the natural photoperiodic variations of their breeding-area. The birds of group 2 (middle diagram) were flown directly into their central African winter range (Zaire 2 ° 14′ S, 28 ° 39′ E); here they were held under natural light conditions, one half of the birds outdoors, the other half indoors. The birds of group 3 (lower diagram) were transferred to chambers with constant temperature and a constant equatorial photoperiod of 12 h. It is clear from Fig. 6.5 that the birds of all three groups behaved very similarly. Following the termination of autumnal zugunruhe, they all underwent a complete winter molt, typical for the species. Subsequently, all birds initiated vernal zugunruhe in February and March, the normal time of northward migration. The close similarity in the timing of vernal zugunruhe between the birds held in constant conditions and those exposed to the natural conditions of central Africa suggests that the onset of vernal migration in the latter group, as well as in free-living conspecifics, was essentially due to the action of a circannual clock, and that environmental stimuli experienced by free-living birds in their equatorial winter quarter have, at most, minor effects.

Results comparable to those shown in Fig. 6.5 have been obtained in other studies with several species of long-distance migrating warblers (Berthold et al. 1972b), flycatchers (Gwinner and Schwabl-Benzinger 1982), and shrikes (Gwinner and Biebach 1977). All show that vernal zugunruhe of birds held in a constant equatorial environment commences at about the normal time. A spontaneous initiation of vernal migratory restlessness under constant conditions has even been observed in birds that normally spend the winter at higher latitudes (e.g., chiffchaff, Gwinner 1971a; whitethroat, Gwinner 1983; blackcap, Berthold et al.

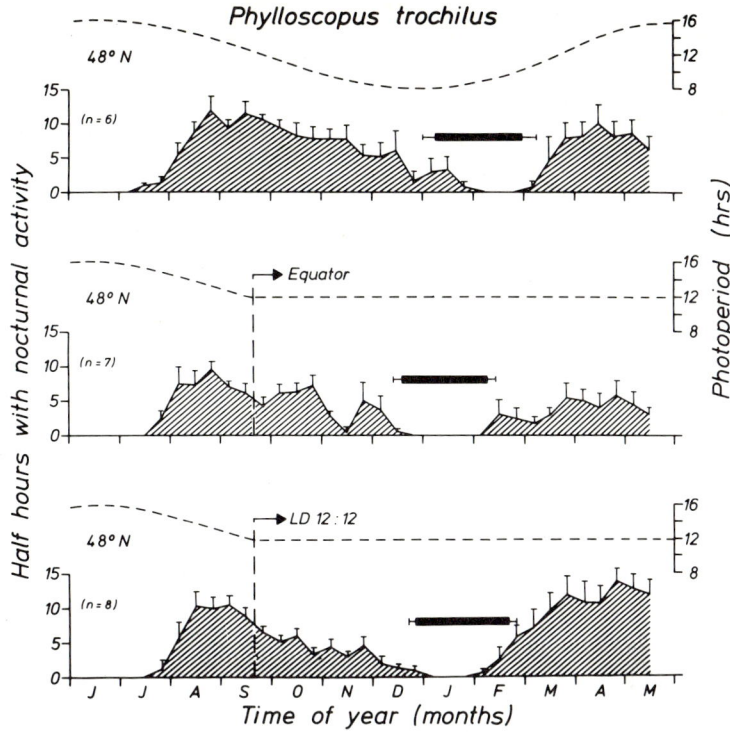

Fig.6.5. Variation in zugunruhe and occurrence of molt of three groups of willow warblers (*Phylloscopus trochilus*). *Upper diagram:* Birds held throughout the experiment in the natural photoperiodic variations of their German breeding grounds at 48 °N. *Middle and lower diagram* birds held until late September in the natural photoperiodic conditions of their breeding grounds. Then the birds in the *middle diagram* were flown to their central African wintering area (Bukavua, Zaire, 2°14' S, 28°39' E) and either held outdoors or indoors under natural photoperiodic conditions. The birds in the *lower diagram* were moved at the same time to constant condition chambers and held there in a constant 12-h photoperiod (LD 12:12) and under constant temperature conditions. ▨ zugunruhe: number of half-hour intervals during which a bird was active each night; mean values for successive thirds of a month are plotted. ■ molt. (After Gwinner 1968a)

1972a; stonechat, E. Gwinner and J. Dittami, unpublished). In these, however, the times of onset of zugunruhe often deviated from the natural schedule, indicating that the appropriate timing is only achieved if the experimental photoperiods match those normally experienced by free-living birds (see Chap. 3.1).

Limited evidence suggests that species differences in the timing of vernal migration and the preceding prenuptial molt may reflect corresponding differences in the onset of actual vernal migration. Thus in a constant 12-h photoperiod, collared flycatchers (*Ficedula albicollis*), that normally spend the winter beyond the equator, molted earlier and began vernal zugunruhe earlier than pied flycatchers (*F. hypoleuca*), that normally winter slightly north of the equator. This presumably reflects differences in the field since it can be assumed that the collared flycatcher begins migration from its more southerly wintering grounds earlier than the pied flycatcher with its more northerly wintering area. Unfortunately there are no rigorous field data in support of this hypothesis (Gwinner 1986).

6.2.2.2 Onset of Autumnal Migration

Like vernal migration, autumnal migration and concomitant events do not necessarily depend on external seasonal variations as they occur in individuals held in constant environmental conditions. This is particularly clear from comparative studies on autumnal zugunruhe of first-year birds. As an example, Fig. 6.6 shows the temporal arrangement of plumage development, body weight, and zugunruhe of first-year willow warblers and chiffchaffs maintained from an age of about

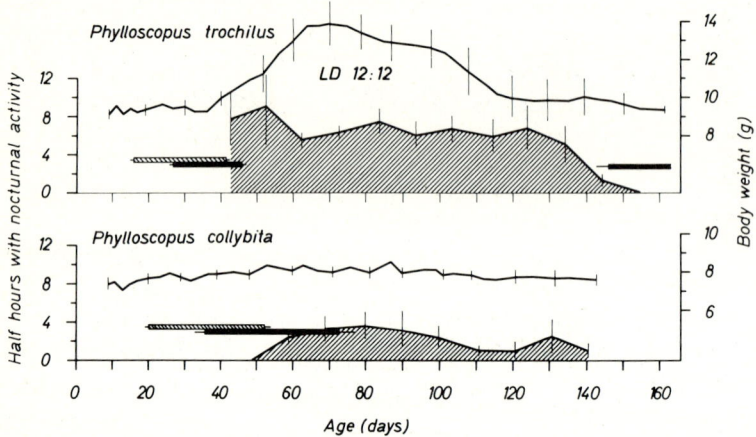

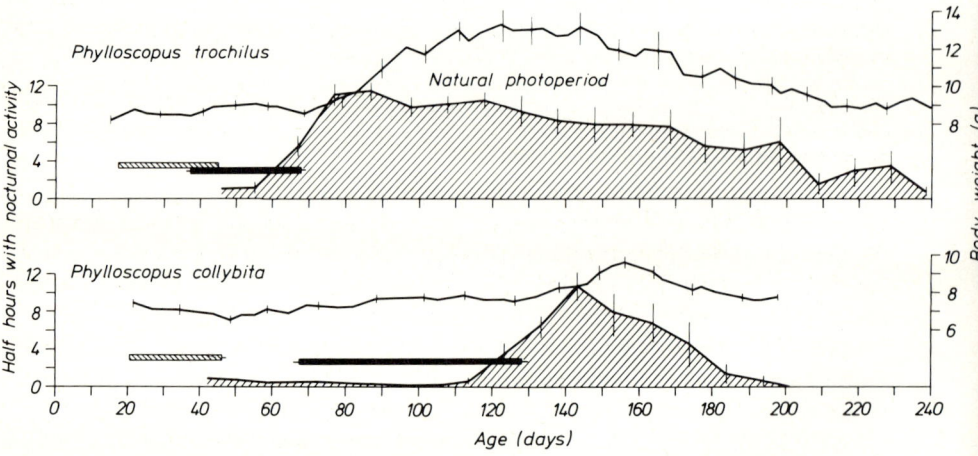

Fig. 6.6. Variations in autumnal zugunruhe ▨ and body weight ——, and occurrence of postjuvenile molt ■ and second phase of plumage development ▧ in willow warblers (*Phylloscopus trochilus*) and chiffchaffs (*Ph. collybita*) raised and subsequently held either under the natural photoperiodic variations in their German breeding areas (*lower figure*) or in a constant 12 h photoperiod (*upper figure*). *Horizontal lines at the end of the bars* and *vertical lines at the curves* represent standard errors. For further explanation see Fig. 6.2. (Gwinner et al. 1971)

9 days under either the natural photoperiodic variations of their breeding area or a constant 12-h photoperiod. Corresponding to their differential migratory and reproductive habits, there are conspicuous differences in the patterns of molt, migratory restlessness and fattening between these two species. Since willow warblers breed later in the year, but leave the breeding area earlier than chiffchaffs, there is not much time available between hatching and autumnal migration (about 12 weeks for willow warblers as compared with about 18 weeks in the chiffchaffs). To be ready for autumnal migration, the willow warblers therefore must accelerate the developmental processes between hatching and departure. Figure 6.6, lower diagram, shows that in birds raised and maintained under the natural photoregime of their breeding grounds, postjuvenile molt started and ended earlier and was shorter in the willow warblers than in the chiffchaffs. Zugunruhe and fattening occurred in both species after the molt was more or less completed, i.e., much earlier in the willow warblers than in the chiffchaffs. In addition to these differences in timing, Fig. 6.6 clearly shows differences between the two species in the duration, amount, and amplitude of autumnal fattening and zugunruhe. All of these parameters were greater in willow warblers as long-distance migrants than in chiffchaffs (see also p. 114).

Theoretically these differences could be the exclusive result of slight differences in the environmental conditions, since willow warblers in average hatch later in the year and are, therefore, exposed to other environmental conditions than chiffchaffs at any given age. They might also be due to differences in the responsiveness of the two species to environmental changes. Figure 6.6, upper diagram, demonstrates, however, that all the essential differences could be observed in birds kept in a constant environment. These birds were taken from the nest at an average age of 9 days and raised under a continuous 12-h photoperiod. There were, it is true, conspicuous differences between these birds and their conspecifics under natural photoperiodic conditions, indicating effects of photoperiod on the endogenous programs. On the other hand, it is clear that environmental factors cannot be responsible for those species differences that persisted in the same constant environment. They must have been due to differences in the endogenous organization of a program that regulates the temporal pattern of the processes investigated.

Even differences between subspecies or populations may be partly due to a differential endogenous programming of seasonal events. This is suggested by the results of an experiment in which willow warblers from northern Sweden (subspecies *acredula*) and from southern Germany (subspecies *trochilus*) were raised and subsequently held in the same constant 18-h photoperiod. *Acredula* showed relatively intense zugunruhe already at an age of 25 days, even before postjuvenile molt had started, and zugunruhe continued throughout the molting period. *Trochilus,* in contrast, did not initiate intense zugunruhe before an age of about 60 days, when postjuvenile molt was almost completed (Gwinner et al. 1972). The difference between the two subspecies in the age of onset of migratory restlessness may be related to the fact that the warblers from the north breed later in the year than those from the south. The beginning of autumnal migration, however, appears to be not correspondingly delayed in the northern subspecies, so that young from the north presumably begin migration at an earlier age than those from the south. This may not leave enough time for them to complete postjuvenile molt before the onset of migration. – Differences similar to those found in captive birds in the timing of zugunruhe relative to postjuvenile molt have been described for free-living willow warblers of the USSR, where Bluymental and Dolnik (1966) found that northern birds from Karelia begin autumnal migration long before the end of postjuvenile molt, whereas more southerly birds from eastern Prussia begin autumnal migration when postjuvenile molt was closer to completion.

Population differences in the patterns of premigratory events of young migrants may also be reflected in birds that are held under a changing photoperiod. There is clear evidence from free-living garden warblers that northern birds from Finland begin migration at an earlier age than southern birds from Germany. Figure 6.7 A shows that a corresponding difference in the timing of zugunruhe

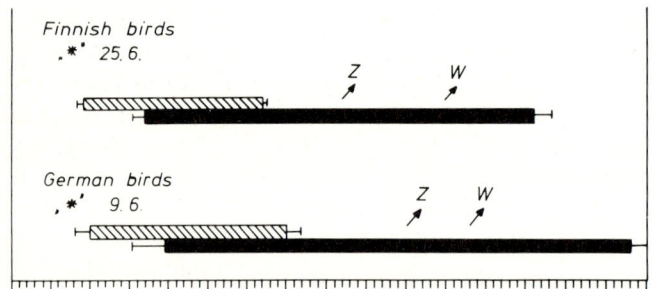

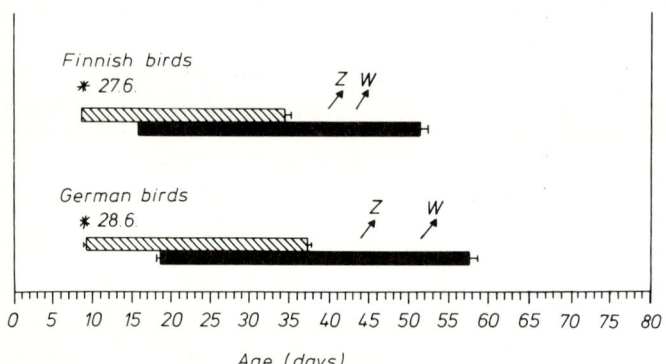

Fig. 6.7. Second phase of plumage development ▨ and postjuvenile molt ■ as well as onset of autumnal zugunruhe ↗^{z} and body weight increase ↗^{w} in Finnish and German garden warblers (*Sylvia borin*). Birds in *A* were raised in the photoperiodic variations of their own respective breeding grounds. Birds in *B* were all raised in the photoperiodic conditions of Germany. * mean hatching date. *Horizontal lines at the bars* represent standard errors. (After Berthold et al. 1974; Gwinner 1979)

was also observed between caged individuals of those two populations. The Finnish birds initiated zugunruhe and fattening, ended their plumage development, and molted earlier than their German conspecifics. This indicates that the entire process of juvenile development proceeded more rapidly in the Finnish birds than in the German birds. The observed differences were at least partly due to inherent differences between the populations as they were also expressed in birds held from an age of 9 days or less in the same photoperiodic variations (Fig. 6.7 B). The differences in the timing of events between the two German populations in A and B are presumably due to the fact that the birds in B hatched about 3 weeks later than those in A. As shown in other studies (see Sect. 6.2.2.6), birds from late broods develop more rapidly than those from earlier broods, as a result of the different photoperiodic conditions to which they are exposed.

Population differences in the temporal pattern of postjuvenile molt preceding autumnal migration have been shown to be under genetic control. Blackcaps (*Sylvia atricapilla*), like other species, obey the rule that postjuvenile molt begins earlier and is of shorter duration in northern populations, which migrate earlier and over longer distances than in southerly populations, which migrate late and over shorter distances, or not at all (e.g., Fig. 6.8, upper diagram). In accordance with the decrease of migratory distance and a progressively later onset of the first migratory period, there was a progressively later onset and shorter duration of postjuvenile molt from Finnish through German and French to Canary Island populations. Figure 6.8, lower diagram, indicates that F1-hybrids bred in aviaries from German and Canary Island parents had an intermediate time course of postjuvenile molt, that was significantly different from that of the parental populations (Berthold and Querner 1982).

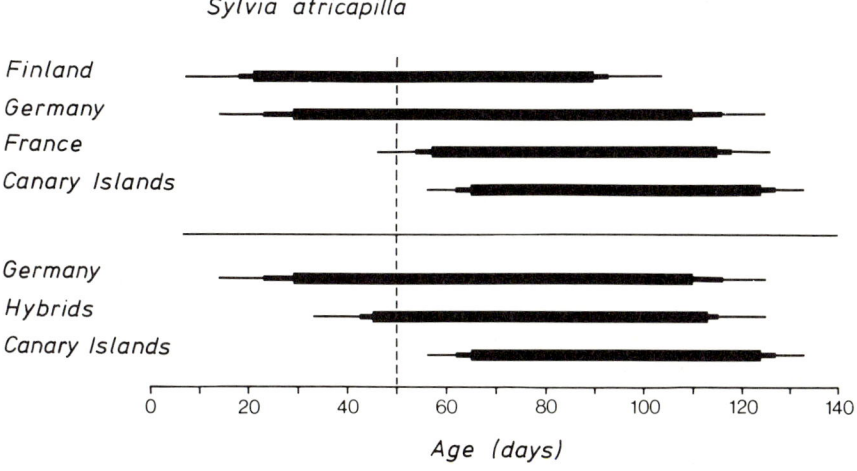

Fig. 6.8. Postjuvenile molt (with standard deviations and standard errors) of blackcaps (*Sylvia atricapilla*). *Upper diagram* results obtained from the 4 populations indicated. *Lower diagram* results obtained from F1-hybrids compared with those obtained from birds of their parental populations. Birds were raised and subsequently held in the natural photoperiodic conditions of southern Germany and moved to a constant 12.5-h photoperiod at the time indicated by the *dashed vertical line*. (After Berthold and Querner 1982)

6.2.2.3 Pattern of Autumnal Migration

In several migratory species, an endogenous circannual rhythmicity not only controls the onset of autumnal and vernal migration and the sequence of associated events, but also the temporal course of migration, at least during the first southward migration of a bird. This was suggested when the temporal variations of autumnal migratory restlessness of caged first-year willow warblers were compared to the temporal variations of actual migration in free-living conspecifics (Gwinner 1968a, b, 1971a). This comparison revealed a striking parallelism between the time course of zugunruhe in caged birds and that of the estimated migratory performance of free-living birds. There was a rapid increase of zugunruhe at the beginning of the migratory season, when free-living conspecifics rapidly increase their rate of travel. The period of most intense zugunruhe coincided with the period during which willow warblers normally cross the Mediterranean and the Sahara desert at their relatively highest rate. Then zugunruhe slowly decreased, in just the same way as free-living birds reduce their rate of migration as they approach their wintering area.

Differences among species in the temporal pattern of autumnal migration have been shown to be reflected, at least grossly, in the pattern of zugunruhe of caged individuals. As shown in Fig. 6.9 chiffchaffs had a shorter period of autumnal migratory restlessness than willow warblers, corresponding to their shorter migratory distance (cf. Fig. 6.1). Their highest levels of zugunruhe were lower than those of willow warblers, consistent with the observation that chiffchaffs presumably never reach the high migratory performances attained by willow warblers during their trans-Sahara migrations. Finally, during the first part of the migratory period, the intensity of zugunruhe increased more slowly in the chiff-

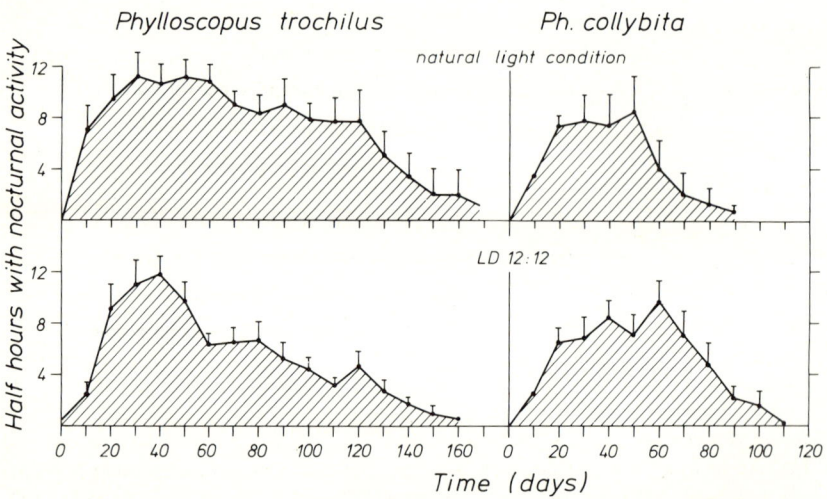

Fig. 6.9. Variations in autumnal zugunruhe in first-year willow warblers (*Phylloscopus trochilus*) and chiffchaffs (*Ph. collybita*) held either under natural photoperiodic variations of southern Germany (*upper two diagrams*) or under a constant 12-h photoperiod (LD 12:12). For other explanations see Fig. 6.2. (After Gwinner 1968b)

chaffs than in the willow warblers, reflecting the more leisurely beginning of migration in the former.

Species differences in the temporal pattern of intensity of zugunruhe in relation to differential migratory patterns of free-living birds have also been found in comparative studies of silviinine warblers (Fig. 6.10). The long- and middle-dis-

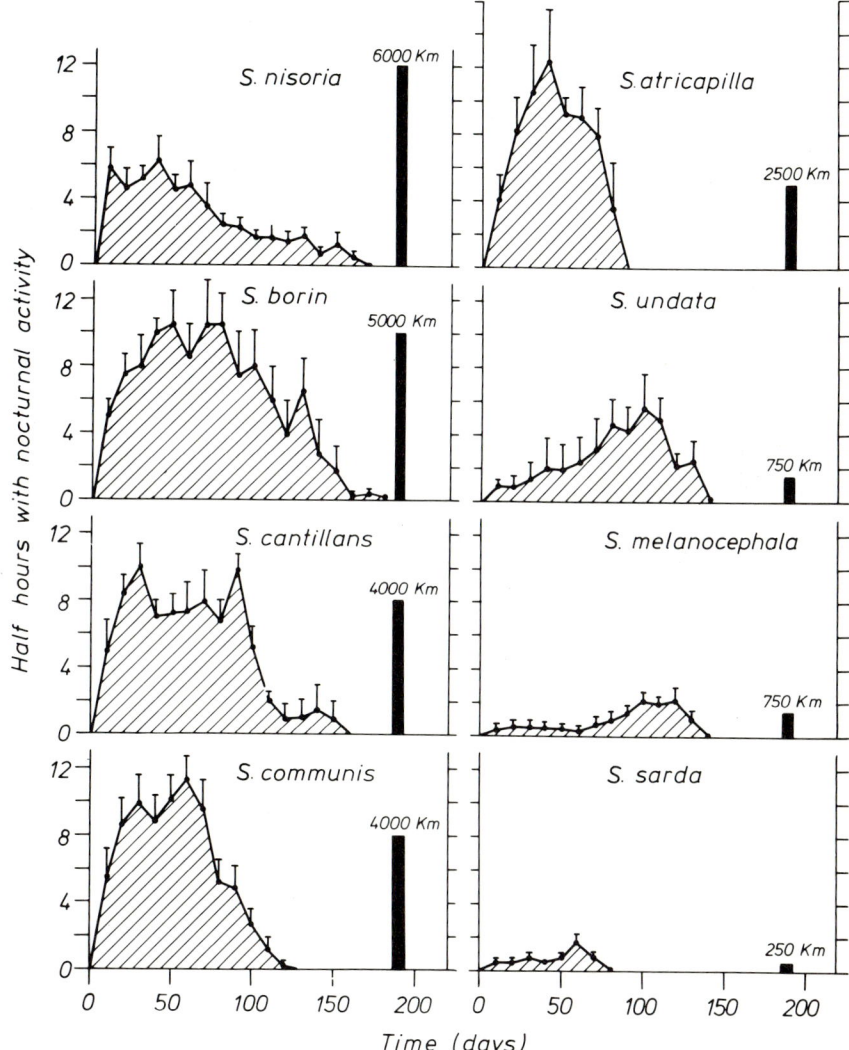

Fig. 6.10. Variations in autumnal zugunruhe in first-year birds of eight species of *Sylvia* (*S. nisoria:* barred warbler; *S. borin:* garden warbler; *S. cantillans:* subalpine warbler; *S. communis:* whitethroat; *S. atricapilla:* blackcap; *S. undata:* Dartford warbler; *S. melanocephala:* Sardinian warbler; *S. sarda:* Marmora warbler). Birds were held under photoperiodic conditions approximating those they would normally have experienced in nature. *Black columns* indicate the approximate migratory distance covered by free-living conspecifics. For other explanations see Fig. 6.2. (After Berthold et al. 1970, 1972a; Berthold 1973b, 1979b; Gwinner 1983)

tance migrants like the garden warbler (*S. borin*), the barred warbler (*S. nisoria*), the subalpine warbler (*S. cantillans*), the whitethroat (*S. communis*) and the blackcap (*S. atricapilla*) that cross the Mediterranean and/or the Sahara on autumnal migration show left-skewed (long-distance migrants) or bell-shaped (middle-distance migrants) patterns of zugunruhe with pronounced peaks early in the season, when the birds normally cross these ecological barriers. In contrast, short-distance migrants, like the Dartford warbler (*S. undata*), the Sardinian warbler (*S. melanocephala*) and the Marmora warbler (*S. sarda*), which are not confronted with especially severe conditions at any place along their migratory route, have rather flat and occasionally right-skewed patterns without any pronounced peaks. The idea that the time course of autumnal zugunruhe is endogenously preprogrammed is also supported by the behavior of species in which an unusual time course is reflected in the pattern of zugunruhe of caged conspecifics. European marsh warblers (*Acrocephalus palustris*), for example, slow down or interrupt autumnal migration for some time in southern Sudan or Ethiopia before proceeding toward their southern African wintering grounds. This reduction of migratory performance is reflected, at least to some extent, by a corresponding reduction of zugunruhe of caged marsh warblers held in an LD 12:12 (Berthold and Leisler 1980).

Like zugunruhe, the variations in body weight, mainly reflecting changes in the amount of fat deposited, have species-specific characteristics related to the birds' migratory habits. As a general rule species migrating over long distances, and thereby usually across severe ecological barriers, deposit more fat and stay longer in a fat condition than species migrating over shorter distances. This is clear in the comparison of the body weight pattern of the long-distance migrating willow warbler with that of the middle-distance migrating chiffchaff (Figs. 6.3 and 6.6) but has also been described for sylviinine warblers (Berthold et al. 1972a), flycatchers (Gwinner and Schwabl-Benzinger 1982), and others (e.g., Odum 1960). Although most of these differences were found in studies in which birds were held under naturally varying photoperiodic conditions, there are several investigations in which corresponding differences could be detected even in birds held under constant conditions, confirming the idea of inherently fixed programs (e.g., Gwinner et al. 1971; Gwinner and Schwabl-Benzinger 1982).

6.2.2.4 Duration of Autumnal Migratory Activity – A Factor Determining Migratory Distance?

The close similarity between the pattern of autumnal migratory restlessness in caged species of *Sylvia* and *Phylloscopus* and that of actual migration in free-living conspecifics has suggested the idea that an endogenous migratory time program may be involved in determining the distance over which a bird travels during its first autumn migratory season. Explicitly, the following hypothesis was proposed (Gwinner 1968a):

1. The zugunruhe of caged birds is the expression of an endogenous time program controlled by circannual rhythmicity that determines duration and temporal variations of migratory activity; and

2. This temporal program is organized in a species- or population-specific fashion in such a way as to produce just enough migratory activity (i.e., migratory time) to reach the specific winter quarters. In other words, the birds whould arrive in the vicinity of the wintering area when the program is terminated.

Such a mechanism could explain a phenomenon so far not properly understood. There is evidence from several experiments suggesting that older birds, at least on their second flight to the wintering quarters, find their wintering ground through goal orientation. When displaced from their migratory route, adult birds compensate for the displacement by changing their angle of migration so that they eventually reach their proper wintering area. Inexperienced first-year birds, in contrast, are apparently only capable of direction orientation. When experimentally displaced, they continue migrating in the original direction and end up in a "wrong" winter quarter (e.g., Drost 1938; Perdeck 1958; Wolff 1970; for reviews see e.g., Gwinner 1971b; Schmidt-Koenig 1975). This raises the question concerning the factors that normally terminate migration in first-year birds, that are apparently capable only to fly in a particular direction. Several mechanisms have been suggested, but so far none have received convincing support (for reviews see e.g., Wallraff 1960; Gwinner 1971b, 1972a, 1977b). The present hypothesis proposes that the first-year birds find their wintering ground through a mechanism of "vector navigation" (Jander 1963) in which the length of the vector is determined by an endogenous circannual program that determines the time that a bird spends in migratory flight.

In the following this hypothesis will be critically evaluated by asking the questions: (1) Are we indeed dealing with a time program? (2) Is this program organized in a species or population specific way? (3) Is the program organized in such a way as required for the bird to find its winter quarters?

1. Time Program?

Several experiments with caged first-year garden warblers indicated that the amount of zugunruhe developed by a bird at a particular time is largely independent of its energetic state and the intensity of zugunruhe developed before that time. Figure 6.11 schematically summarizes the results of four of these experiments. In (a) an experimental group of warblers was prevented from depositing fat at the beginning of the autumnal migratory season by limiting the food uptake. The pattern of zugunruhe was not affected by this treatment as revealed by the comparison with the pattern of the control birds that had received food ad libitum throughout. The reduction of body weight by controlled starvation to a level just above the lean weight later in the migratory season also had no effect (b). A drastic reduction of zugunruhe occurred only when food was limited to an extent that body weights reached levels characteristic for the complete exhaustion of fat reserves (c). When these birds were returned to food ad libitum both body weight and zugunruhe recovered to control levels. There was no compensation for the previous zugunruhe gap, so that the experimental birds in total showed less zugunruhe than the controls. The same was true for a group of warblers (d) whose zugunruhe was suppressed for 2 months by exposure to complete darkness at night: after these birds were returned to dim night light, their zugunruhe pattern

Sylvia borin

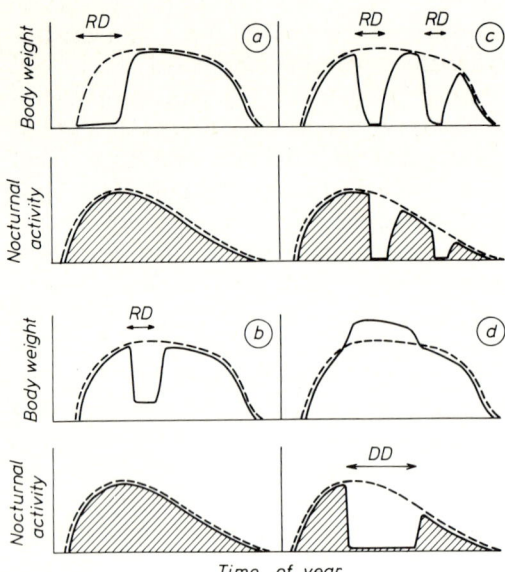

Fig. 6.11. Schematic representation of four experiments with garden warblers (*Sylvia borin*). *Broken curves* behavior of control birds; *solid curve* behavior of experimental birds. The experimental birds of *a, b,* and *c* received a reduced diet (*RD*) and those of *d* were exposed to complete darkness at night (*DD*) during the times indicated. For further explanation see text. [After Berthold 1977a (*a*), 1975a (*b*), 1976b (*c*); Gwinner 1974b (*d*)]

was identical to that of the controls, despite the fact that they had increased body weights during the time of reduced nocturnal activity, suggesting a decreased energy expenditure.

These and similar results are consistent with the idea that the pattern of zugunruhe is primarily the result of an endogenous time program that determines, phase by phase, the intensity of migratory activity. Although the overt expression of this program can be affected by certain manipulations like exposure to complete darkness (d), drastic reduction of body weight below fat-free level (c) or refeeding following starvation (Biebach 1985; Gwinner et al. 1985a), the underlying temporal mechanisms seemed to remain uninfluenced as zugunruhe always returned to control levels after conditions were brought back to normal. The temporal pattern of zugunruhe seems to be determined by a continuously changing setpoint that determines migratory time, and that is shifted according to a predetermined temporal course by a circannual clock.

2. Species and Population Differences?

The hypothesis predicts that the time program for migration differs among species or populations according to differences in their migratory distance. This prediction was verified in comparative studies with species of *Sylvia* and *Phylloscopus* that travel over different distances during migration (Figs. 6.9 and 6.10): The total amount of time spent by a species in autumnal migratory restlessness was proportional to the distance between breeding grounds and wintering quarters.

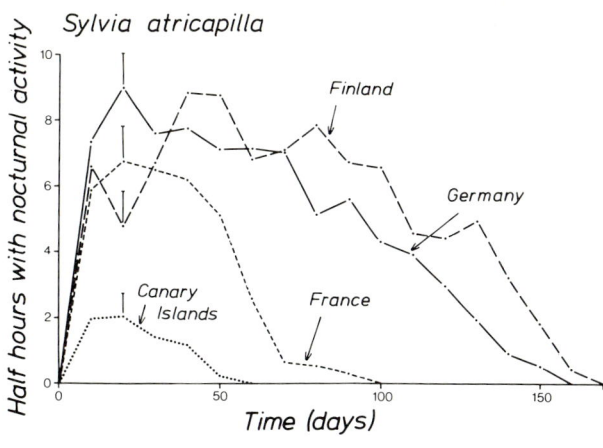

Fig. 6.12. Variations of autumnal zugunruhe in four populations of blackcaps (*Sylvia atricapilla*) held during their first 50 days of life under the natural photoperiodic conditions of Germany and then in a constant 12.5-h photoperiod. For further explanation see Fig. 6.2. (After Berthold and Querner 1981)

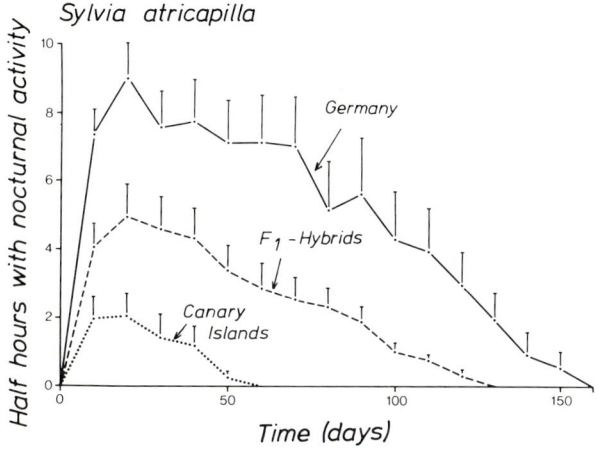

Fig. 6.13. Variations of autumnal zugunruhe of F1-hybrids between blackcaps (*Sylvia atricapilla*) from Germany and the Canary Islands, compared with that of the parental stock. For further explanation see Fig. 6.2. (After Berthold and Querner 1981)

Even populations of the same species that differ with regard to their migratory distances show different quantities of autumnal zugunruhe under identical experimental conditions. Figure 6.12 shows autumnal migratory restlessness curves of blackcaps from four different populations, Finland, southern Germany, southern France, and the Canary Islands. Finnish birds are long-distance migrants, travelling far into Africa; under experimental conditions they showed the most intense nocturnal activity. The German, French, and Canary Island populations have progressively shorter migratory distances and correspondingly developed progressively less zugunruhe under caged conditions.

That these population differences in blackcaps are under genetic control is clear from the behavior of hybrids (Fig. 6.13). F1-hybrids obtained from southwest German and Canary Island parents developed a pattern of autumnal zugunruhe intermediary between that of the respective parental population. Such an intermediary behavior of the F1-hybrids would be expected under the likely assumption that zugunruhe is controlled by many genes.

3. Control of Distance

The hypothesis assumes that the temporal program controlling migration produces just enough migratory activity during the first fall for the birds to reach their specific wintering quarters. This central prediction of the hypothesis proves to be extremely difficult to test and still lacks confirmation.

An obvious way of testing it would be to multiply the number of hours during which a bird showed migratory restlessness in the cage with the average speed at which it would normally migrate. If the hypothesis were correct, the distance thus calculated and the actual distance covered by free-living individuals should be identical.

Unfortunately, this approach is difficult for at least two reasons. First, the average speed of the migratory flight is not known for any of the species studied, and second, the actual distance covered by free-living individuals of the species studied is not known either and, indeed, is very difficult to obtain. Although we can calculate the direct distance between the breeding and the wintering area, the actual distance travelled in migration must be considerably longer, as is known from radar studies and direct observations. Migration does not proceed along straight lines but rather in a zig-zag fashion. The question of how much this zig-zagging adds, on the average, to the direct distance, cannot even be reasonably estimated.

As a consequence of these difficulties, the prediction can only be tested in an indirect way. One such approach is based on the comparison of known migratory performances of free-living birds with the amount of zugunruhe (expressed as the total number of half-hour intervals with nocturnal activity) displayed by caged birds at corresponding intervals (Fig. 6.14). The hypothesis assumes that the amount of zugunruhe (U_i) shown by a caged bird within a certain time interval is equivalent to a distance D_i travelled during this interval by a free-living conspecific. The ratio $\frac{U_i}{D_i}$ should then equal the ratio $\frac{U_g}{D_g}$, where U_g is the total amount of zugunruhe of the caged bird, and D_g the unknown total distance travelled by the free-living birds. It can be calculated as $D_g = U_g \cdot \frac{D_i}{U_i}$ and if the

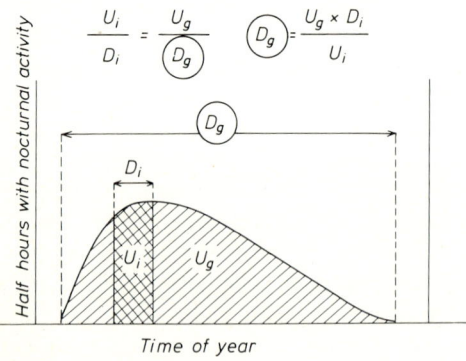

Fig. 6.14. Schematic representation of autumnal zugunruhe in first-year willow warblers (*Phylloscopus trochilus*). For further explanation see text

hypothesis proposed here is correct, the calculated distance should be approximately that normally covered by the population during migration.

Fortunately, information about D_i is available from recoveries of banded birds. To test the hypothesis, data from 24 willow warblers and 20 chiffchaffs, banded during autumnal migration in northern Europe and recovered during the same migratory season in southern Europe, were used to determine D_i for 24 and 20 individual birds respectively. Using the formula above, mean D_g for individuals were calculated from D_i for 21 caged willow warblers and 14 caged chiffchaffs with known U_i and U_g. These means of D_g values were averaged again for the 6 to 8 individuals of three experimental groups of willow warblers and two experimental groups of chiffchaffs to which these birds belonged. Then the endpoints of the route along which each experimental group would have flown were plotted on a map, assuming that the birds had travelled along the normal flyway.

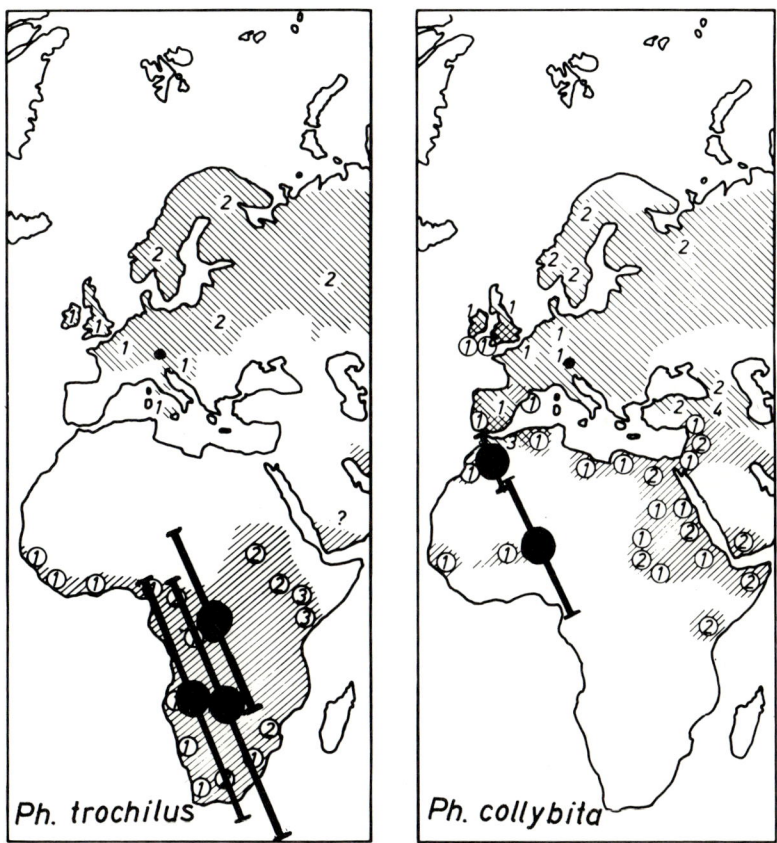

Fig. 6.15. Breeding areas and winter quarters of the willow warbler (*Phylloscopus trochilus*) and the chiffchaff (*Ph. collybita*). *Numbers* refer to breeding and wintering areas of various subspecies. *Large solid circles* are calculated endpoints of migration (with standard errors) for two groups of chiffchaffs and three groups of willow warblers kept in slightly different experimental conditions. For further explanations see text. (After Gwinner 1972a)

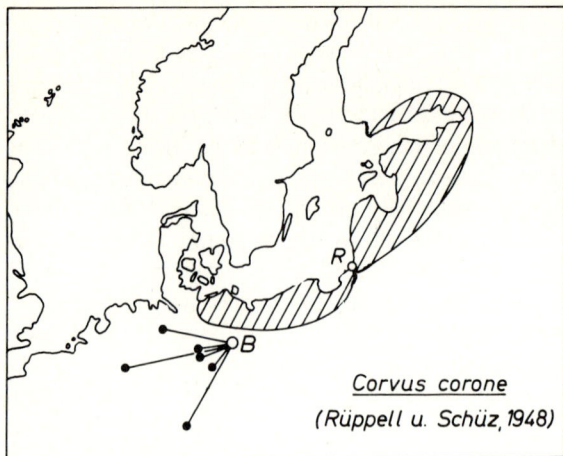

Fig. 6.16. Winter recoveries of carrion crows (*Corvus corone*) that were caught on their autumnal migration in Rossitten (*R*) and released in Berlin (*B*) immediately after capture. *Hatched area* main breeding and wintering area of the population studied. (After Rüppell and Schüz 1948)

These calculations suggest (Fig. 6.15) that the birds would, in fact, have reached their respective wintering areas.

Although the results of these calculations are consistent with the proposed mechanism, they are certainly not sufficient for the final evaluation of the present hypothesis. More critical experimental data are clearly desirable. The difficulties mentioned above can probably not be resolved by further laboratory experiments. On the other hand, field experiments with migrating birds may be rewarding. Displacement or detention experiments could be especially helpful in solving the problem, as the present hypothesis makes strong predictions about the behavior in these situations. An example of such an experiment is shown in Fig. 6.16, which summarizes results obtained with carrion crows (*Corvus corone*). First-year birds were caught in fall, halfway along their migratory route, and displaced to a location beyond their normal wintering area. Winter recoveries revealed that the displaced individuals had continued to migrate in their original direction, and over distances similar to that which had separated the birds from their actual wintering area at the point of capture.

Another displacement experiment, carried out by Perdeck (1964) with first-year European starlings, also produced results consistent with the present hypothesis, although they indicated at the same time that the actual duration of autumnal migration was also subject to environmental modification, at least toward the end of the migratory season (Fig. 6.17). Starlings were caught in Holland (X in Fig. 6.17) on their autumnal migration from the Baltic area to Holland, Belgium and southern England, and displaced southward to Barcelona (X') where they were banded and released. The displaced birds were from two populations, as known from previous investigations of banded birds: a population with a more westerly breeding (A) and wintering (A') distribution that migrates early through Holland and that is still some distance from its wintering area when caught in Holland; and another population with a more easterly breeding (B) and wintering (B') distribution, that passes later through Holland and that has almost reached its wintering area upon capture. After release the young birds continued to travel

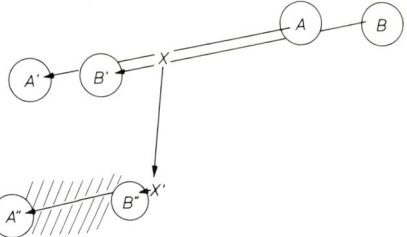

Fig. 6.17. Schematic representation of a displacement experiment of Perdeck (1964) with European starlings (*Sturnus vulgaris*). *Hatched area* unsuitable area for wintering. For further explanations see text

in their original west-southwesterly direction, consistent with previous results on this species (p. 117). In doing so, however, the birds reached areas suitable for wintering only either close to the release sites (in the Ebro valley) or far away in the southwestern part of the country (Estramadura and the valley of Guadalquivir). Banding recoveries of wintering birds revealed that most birds from population A spent the winter in the latter area (A″), i.e., they travelled over distances even longer than normal for this population in autumnal migration. In contrast, most of the birds from population B remained in the favorable vicinity of Barcelona (B″), so that their migratory distance was shorter than that normal for this population. The results are best interpreted by a hypothesis that both endogenous temporal factors and external condition determine how far a bird migrates: At the beginning of the migratory season the internal "drive" dominates; thus birds from population A continued to migrate over a long distance following release in Spain. At the end of the season, in contrast, external factors dominate; thus the birds from population B remained in the area of Barcelona with its favorable conditions, and hence migrated over a shorter distance than normal.

The hypothesis also predicts that detaining a bird for some time during its autumnal migration, and then releasing it, should cause it to winter closer to its home area. Results of Bellrose (1958) with blue-winged teal (*Anas discors*) are in agreement with this prediction. Teal were caught in Illinois on their first autumnal migration and detained until all the adults had passed through. When released they continued in the appropriate direction, but winter recoveries of the detained birds were, on the average, at shorter distances from the ringing site than recoveries of control birds that had been released immediately. While about 50% of the latter were recovered at distances greater than 2000 km from the release site, only about 10% of the detained birds were found that far away. Although other interpretations are possible, these results can obviously be easily accommodated by the hypothesis.

4. Critique of the Hypothesis

In the previous section, emphasis has been placed on evidence supporting the hypothesis that first-year migrants reach their wintering area by migrating in a given direction for a given time. There are, however, some data that speak against this model, or, at least, suggest caution as to its general applicability (for a more detailed account see Gwinner 1972a, 1977b, 1986).

First of all it must be mentioned that there are a few cases in which the general rules and predictions discussed in (2) are not sustained. The most obvious instance emerges from a comparative study of the pied flycatcher with its sibling species, the collared flycatcher (Gwinner and Schwabl-Benzinger 1982). The latter species has a longer migratory route than the former, but in three replicate experiments it was always the pied flycatcher, with the shorter migratory distance, that showed a larger overall amount of zugunruhe. A similar discrepancy was found in a study of red-backed shrikes (*Lanius collurio*) from Finland and from southern France (Gwinner and Biebach 1977). Although the Finnish birds migrate further than the French birds, the patterns of autumnal zugunruhe observed in captive birds of these two populations were indistinguishable. In observations on sylviinine warblers (Fig. 6.10), the barred warbler (*S. nisoria*) provides an exception in that it showed relatively little zugunruhe, despite the long distance it migrates.

The patterns of zugunruhe do not always clearly match actual migratory pattern. Discrepancies are found, for instance, among the various populations of blackcaps (Fig. 6.12). Although it is true that the overall amounts of zugunruhe developed by the four groups differed according to the distance covered by the populations from which they were taken, the detailed patterns differed from the expectation: the high maximal values exhibited by the birds from the southern (short-distance migrating) populations are certainly not matched by corresponding maxima in migratory performance of free-living birds of these populations.

None of these apparent discrepancies provide conclusive evidence against the model proposed here, as methodological deficiencies or particularities in behavior may account for them (see e.g., Gwinner 1986). However, they should be taken as a warning against the premature acceptance of its generality.

The most obvious general objection to the hypothetical mechanism proposed here would be its inaccuracy. If migratory distance were merely a passive consequence of the time spent for the migratory flight, the overall distance would depend drastically on environmental conditions, especially weather. For instance, a bird flying with tail winds would cover a much grater distance per time unit than a bird flying in head winds. Even if the migratory time program were adjusted to average meteorological conditions, considerable seasonal and year-to-year variations would still be expected in migratory distance, and, correspondingly, in the exact location of the wintering area. For species such as the willow warbler or the garden warbler with wintering grounds that extend over a very wide range of latitudes, these objections are perhaps not serious, since it is possible that even birds of the same population spend the winter hundreds and perhaps even thousands of kilometers apart. However, the low precision of the hypothetical mechanism would be a problem for species with wintering areas that encompass only a narrow latitudinal range as in the case of the orphean warbler (*S. hortensis*) with a north-south extension of only a few hundred kilometers.

Theoretically it is possible that precision may be improved by internal compensatory processes that reduce the effects of external disturbances, for instance by allowing birds to make up for delays caused by adverse weather conditions with additional flight time. Although some data can be interpreted in terms of such a compensatory mechanism, there are others that clearly speak against such a possibility (see Gwinner 1977b for discussion).

These and other considerations make it likely that the mechanism proposed here is normally modified by external factors, at least during the final part of migration. For short-distance migrants the significance of external cues for the termination of autumnal migration has already been demonstrated, e.g., in the European starling (Fig. 6.17). Moreover, it has been found that autumnal migratory restlessness in several short-distance migrants not only exceeded the time normally spent for migration (Weise 1963; Mewaldt et al. 1964; Gwinner 1972a), but also showed more intra-individual variability (Gwinner 1968b; Berthold et al. 1972a) and appeared to be more susceptible to exogenous variables than in long-distance migrants of the same genus (Palmgren 1937; Wagner and Schildmacher 1937; Wagner 1957; Czeschlik 1976). The relative significance of internal and external mechanisms controlling migratory activity may then differ between long- and short-distance migrants. In the former, internal mechanisms may be of prime significance; in the latter, external factors may dominate. Hence migratory restlessness in many short-distance migrants may express only vaguely the general readiness of a bird to migrate; the actual time course and the distance of migration would then be largely determined by external factors.

6.2.2.5 Migratory Direction

Many migratory birds reach their wintering grounds not by a direct flight but rather by changing direction once or several times en route. Thus, European long-distance migrants that winter in central or southern Africa often leave their breeding grounds in a southwesterly or southeasterly direction and later change to southeast or southwest, respectively (e.g., Zink 1973–1981, 1977). It has been proposed that these shifts in migratory direction might be induced by external (e.g., celestial) cues (e.g., Sauer 1957) or by maintaining a constant angle relative to the dip of the earth magnetic field (Kiepenheuer 1984), but convincing evidence for these hypotheses is lacking. The results summarized in Fig. 6.18 suggest that the directional changes performed by first-year garden warblers may be the result of endogenous circannual variations in the physiological state. In these experiments garden warblers were maintained during their first year of life in a constant 12-h photoperiod and tested repeatedly in orientation cages for directional preferences of their nocturnal migratory unrest. The birds had no view of the sky but were exposed to the natural earth magnetic field. Figure 6.18 shows that the birds changed their directional preferences correspondingly with changes in the migratory direction of free-living conspecifics. Hence, these results are consistent with the idea that an endogenous circannual time program is involved not only in determining the temporal pattern and possibly the distance of migration, but also the seasonal variations in migratory direction, possibly by changing the birds' angle of orientation relative to the earth magnetic field (Gwinner and Wiltschko 1978, 1980).

6.2.2.6 Photoperiodic Modification of the Program

In most of the laboratory studies discussed so far, experimental birds were either held in a constant photoperiod or in photoperiodic conditions simulating those

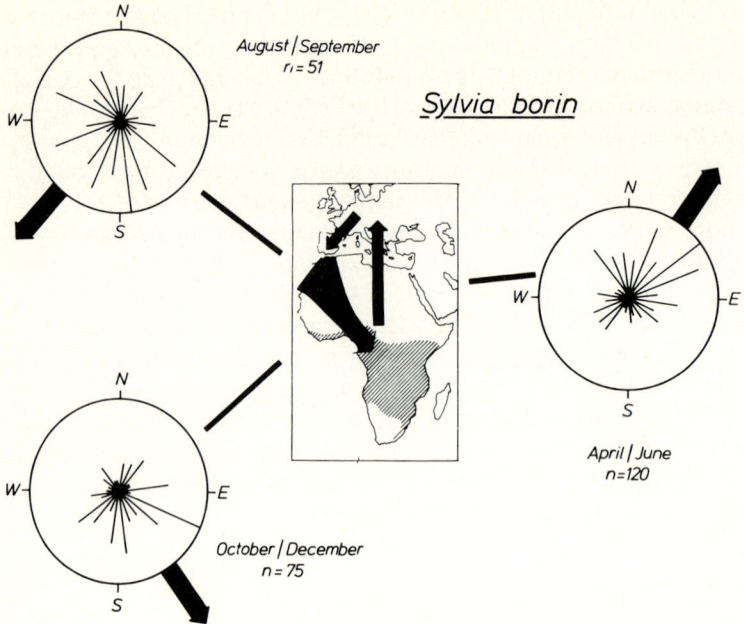

Fig. 6.18. Spontaneous seasonal changes of directional preferences in garden warblers (*Sylvia borin*) during nocturnal migratory restlessness. Garden warblers were kept throughout the experiment under a constant 12-h photoperiod and tested repeatedly in circular orientation cages. The heading each bird assumed during a test night was taken as statistical unit for data analysis. The *three circular diagrams* summarize the results obtained in August and September (*upper left*), in October through December (*lower left*) and in April through June of the following year (*right*). The data are plotted on a relative scale such that the radii equal the greatest amount of activity in any one 15° sector. The *large arrows* at the periphery of each diagram show the direction of the mean vector calculated for each test series. *Numbers* at each diagram refer to the number of tests. The *map* in the center shows schematically the changes in migratory direction known to occur in free-living garden warblers in the course of the year. (After Gwinner and Wiltschko 1978, 1980)

they would experience during migration; little attention has been given to the possible modifying effect of photoperiod. Yet, since photoperiod can act as a zeitgeber for endogenous circannual rhythms (Chap. 4), it should be expected that it also affects the migratory program. The small amount of information available indicates that this is indeed the case and, moreover, that photoperiodic influences have an adaptive, biologically significant character.

It has been observed in several species of sylviinine warblers and in other species that short photoperiods accelerate seasonal processes in summer and early autumn. An example of this is shown in Fig. 6.19. Fall migratory restlessness began much earlier, and the preceding processes of plumage development proceeded much more rapidly in willow warblers kept in a 12-h photoperiod than in those kept in an 18-h photoperiod. This accelerating effect of shorter photoperiods on the timing of the preparation for, and onset of autumnal zugunruhe, is probably advantageous for individuals hatched late in the season and thus growing up under decreasing daylengths. To avoid unfavorable environmental conditions in

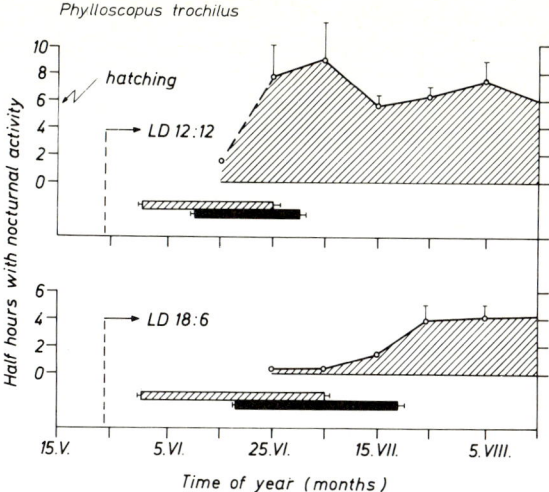

Fig. 6.19. Fall zugunruhe (*cross-hatched*) second phase of plumage development (*cross-hatched bars*) and postjuvenile molt (*black bars*) of two groups of willow warblers (*Phylloscopus trochilus*) transferred at an average age of 9 days (May 24) from the natural photoperiod of their breeding area to constant 12-h (LD 12:12) or 18-h (LD 18:6) photoregimes. *Vertical lines at the symbols* and *horizontal lines at the bars* standard errors. For further explanations see Fig. 6.2. (After Gwinner et al. 1971)

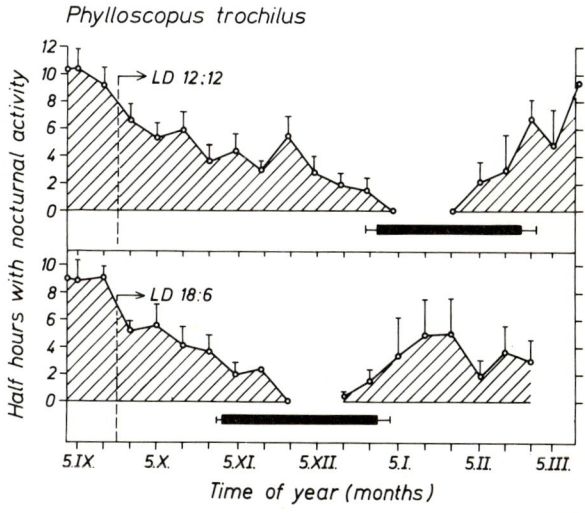

Fig. 6.20. Fall zugunruhe (*cross-hatched*) and prenuptial molt (*black bars*) of two groups of willow warblers (*Phylloscopus trochilus*) transferred at an average age of 110 days (September 20) from the natural photoperiodic conditions of their breeding area to a constant 12-h (LD 12:12) or 18-h (LD 18:6) photoperiod. For further explanations see Figs. 6.2 and 6.19. (After Gwinner 1971a)

fall, these birds initiate migration at an earlier age than those hatched earlier in the year. Consequently, they must pass more rapidly through the preparation of southward migration. The results shown in Fig 6.19, as well as similar results obtained from garden warblers and blackcaps (Berthold et al. 1970), suggest that the differences between birds hatched early and late are at least partly due to the differences in photoperiod during development.

In summer and early autumn it is the shorter photoperiod that accelerates seasonal activities, but the opposite holds true in late autumn, winter and early spring, as can be seen in Fig. 6.20. The onset and end of molt and the onset of vernal migratory restlessness occurred earlier in birds transferred in mid-September to an 18-h photoperiod than in birds transferred to a 12-h photoperiod. This accelerating effect of long photoperiods in winter may be advantageous for those willow warblers that cross the equator on their autumnal migration. Because in these regions daylength increases with latitude during the northern winter, birds migrating far south are exposed to longer photoperiods than those conspecifics that winter further north. As a result, autumnal migratory restlessness may be curtailed and the processes preceding vernal migration may be accelerated. This, in turn, would enable an earlier onset of vernal migration. An earlier onset of migration in birds wintering further south is known for a variety of species (e.g., Curry-Lindahl 1963).

Photoperiodic effects similar to those described above for the willow warbler have also been shown in a few other species, mainly sylviinine warblers (Berthold et al. 1970, 1972a). The way in which photoperiod modifies the endogenous pattern has not as yet been investigated in detail. It is clear, however, that the final evaluation of the model of an endogenous timing of migration will depend on detailed information about the modifying effects of photoperiod.

6.3 Conclusions and Perspective

Pittendrigh (1981a) has emphasized that circadian rhythms serve as clocks in two ways: by "assuring an appropriately stable temporal sequence in the program's successive events, they, in effect, measure the lapse of (siderial) time"; and "by providing for a proper phasing of the program to the cycle of environmental change, they, in effect, recognize local time." As has been illustrated in this chapter, both of these functions are also exerted by many circannual rhythms, justifying their designation as circannual clocks.

Since the period of circannual rhythms deviates from 1 year, a proper phasing must be accomplished between circannual programs and seasonal variations in the environment. This can only be achieved through the interaction of the endogenous rhythmicity with environmental zeitgebers, which adjust both the period and the phase of the "circa"-annual rhythm to the natural year. The consequence of this situation is that any answer to questions concerning the adaptive value of circannual rhythms depends on the complete understanding of these interrelationships. An example illustrating the adaptive responsiveness of a circannual system to zeitgeber stimuli is provided by the photoperiodic modification of seasonal patterns in migratory warblers (Sect. 6.2.2.6). So far, little attention has been given to the dependence of circannual programs on the (variable or constant) properties of environmental variables. Future studies aimed at understanding the adaptive significance of circannual rhythms must take these complexities into account.

Environmental factors not only provide for the appropriate control of period and phase of circannual rhythms, but also set limits to their expression. It was previously suggested for both circadian and circannual rhythms that the factors that are used for the synchronization of a rhythm are also those capable of arresting it once they reach certain values (e.g., light intensity with circadian rhythms, and photoperiod with circannual rhythms; Gwinner and Dittami 1986). This phenomenon is by no means fully understood, but in the field of circannual rhythms there are at least some suggestions about its functional meaning. In the pied flycatcher, an equatorial migrant, the circannual rhythms of reproduction and migration become arrested in mid-winter, if photoperiod is slightly longer than that normally experienced. In the closely related collared flycatcher, in contrast, these longer photoperiods are still permissive for the continuation of rhythmicity. This species winters even further south, beyond the equator, where winter photoperiod is longer than that prevailing in the wintering area of the pied flycatcher. Its tolerance of longer photoperiods can be interpreted as an adaptation to the conditions of the more southerly wintering quarter (Chap. 3.1). Comparative studies of this kind may provide further insight into the question of why the interaction of particular circannual systems with their zeitgebers is organized the way it is.

Since, as has been shown, circannual rhythms exert their function only through the interaction with seasonal environmental variables, the question addressed in the introduction of this chapter must be asked again: for what function have organisms evolved circannual clocks to begin with, if environmental zeitgebers are required anyhow for their synchronization? Possible answers to this question have been suggested in this chapter: some of the properties of circannual rhythms may make them better suited for the control of annual cycles in some organisms than other conceivable control systems (see Chap. 6.1). But because none of these propositions has yet received sufficient support, more investigations are required to settle this basic problem of circannual rhythms research.

Appendix: General Oscillator Model and Terminology

Circannual rhythms, like circadian rhythms, have been shown to continue for many cycles even in animals isolated from periodic changes in the environment. In this respect they behave like physical oscillators in the absence of periodic driving agents and therefore can be described as autonomic oscillating systems.

Figure A.1 illustrates the terms commonly used to describe an oscillating system. It is characterized by its period τ which is the time required for the completion of one cycle. The reciprocal of τ is frequency $\frac{1}{\tau}$. The instantaneous state of the oscillation at any particular time of the period is referred to as its phase. The value on the abscissa corresponding to each phase is the phase-angle (φ_1, φ_2, φ_3 in Fig. A.1). The arithmetic means of all phases within one period is the mean value, or the level, of the oscillation. The difference between the maximal and the minimal value of the oscillation is called the range of oscillation, and the difference between the maximal (or the minimal) value and the mean values is called the amplitude.

Under constant conditions the system oscillates with a period typical for the system, i.e., with its natural period τ_n. The oscillator is then said to be free-running. If the range of the oscillation decreases with time it is said to be damped, if not, it is self-sustained. A free-running oscillator can be reset (displaced along the time axis) by appropriate external forces, i.e., it can be phase-shifted by $\Delta\varphi$. A phase-response curve describes the amount and direction of the shift of the phase angle as a function of the phase at which the stimulus is applied. During a phase shift the oscillation usually requires several cycles to regain a new steady state. Cycles between two steady states are called transient cycles.

The period τ_n of an oscillator can be modified by appropriate periodic forces such that it assumes the period T of the driver (Fig. A.2). The forced, or driven, oscillator is then said to be synchronized with the forcing, or driving, oscillation. If both the driving and the driven system are self-sustained the more general term "synchronization" can be replaced by "entrainment". In that case the driving os-

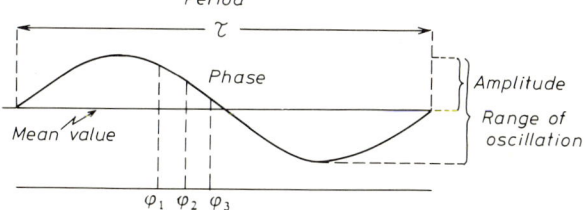

Fig. A.1. Oscillator terminology illustrated with a sinusoidal oscillation. See text

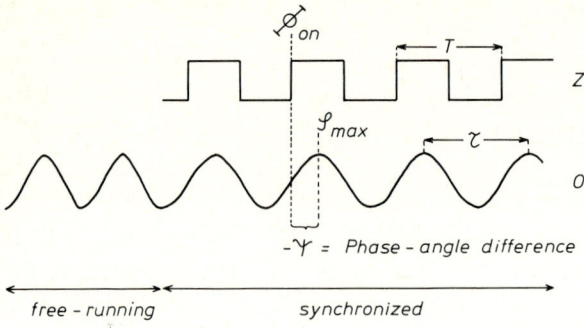

Fig. A.2. An oscillator (sine-curve, O) which is first free-running (*left*) is synchronized by a zeitgeber (*square curve, Z*) the period T of which is longer than the initial free-running period of the oscillator. During entrainment $\tau = T$. The phase angle difference Ψ measured between Φ_{on} and φ_{max} is negative

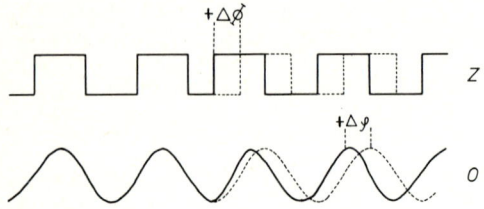

Fig. A.3. An advance phase shift of the zeitgeber (Z) by $+\Delta\Phi$ is followed through one transient cycle by a corresponding phase shift of $+\Delta\varphi$ of the driven oscillator (O). For comparison, *dashed curves* indicate the situation without phase shift

cillation may be called "entraining agent" or "zeitgeber". Entrainment is realized if $\tau_n = T^1$

A phase shift (i.e., a single displacement along the time axis) of the driving oscillator by $\Delta\Phi$ is usually followed by a corresponding phase shift $\Delta\varphi$ of the driven oscillator (Fig. A.3). Delay phase shifts are designated as negative and advance phase shifts as positive. Re-entrainment may be achieved instantaneously or through one to several transient cycles (one transient cycle in Fig. A.3).

Damped and self-sustained oscillations can be synchronized by a driving oscillator only within a certain range of periods (Fig. A.4). Outside this range of entrainment the driven oscillation free-runs[2]. The size of the range of entrainment increases with the strength of the driving oscillator and decreases with the degree of self-sustainment in the driven system.

Within the range of entrainment the driving and the driven systems assume a characteristic phase relationship, the phase-angle difference Ψ (Fig. A.2). To calculate Ψ, a phase angle of the driving and of the driven oscillator must be arbitrarily defined (e.g., Φ_{on} and φ_{max} in Fig. A.2). If the phase angle of the driving oscillator is called Φ and that of the driven oscillator φ, then $\psi = \Phi - \varphi$. Con-

---

[1] The terms entrainment, entraining agent, range of entrainment and zeitgeber are often used in a looser sense, i.e., in connection with the synchronization of both self-sustained and damped systems. Due to the limited number of circannual cycles that can possibly be measured within the life span of an animal the question of whether circannual rhythms are truly self-sustained or damped cannot be answered with certainty. The use of the above-mentioned terms in connection with circannual rhythms can therefore not be taken to imply that these systems are truly self-sustained; it rather follows the flexible application of these terms.

[2] There may be secondary ranges of entrainment around integral multiples or submultiples of the zeitgeber period. This situation is referred to as synchronization by frequency multiplication or demultiplication.

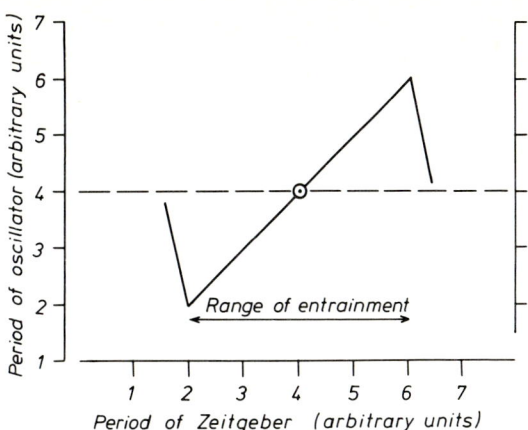

Fig. A.4. Illustration of the range of entrainment. The period of the driven oscillator follows changes of the period of zeitgeber only within certain limits, i.e., within the range of zeitgeber periods of 2 to 6 arbitrary units

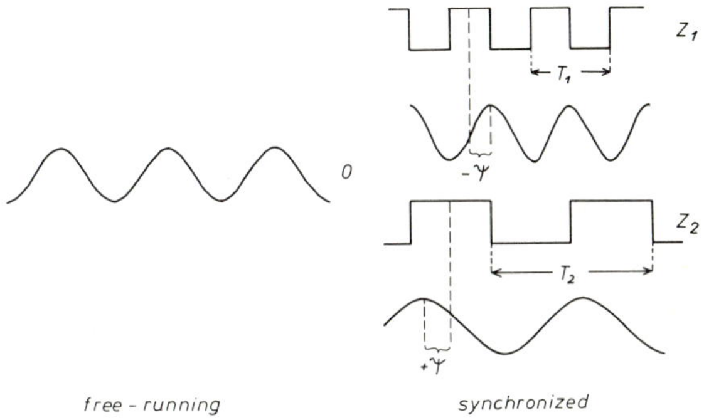

Fig. A.5. An oscillator that first free-runs under constant conditions is synchronized to a zeitgeber with either a short (Z_1, T_1) or a long period (Z_2; T_2.) As predicted, the phase-angle difference Ψ is smaller during entrainment with Z_1 (negative in the example shown) than with Z_2 (positive in the example shown)

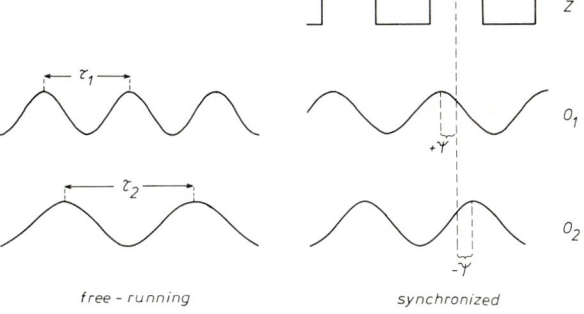

Fig. A.6. An oscillator O_1 with a short natural period τ_1 and an oscillator O_2 with a long natural period τ_2 are synchronized to the same zeitgeber Z. As predicted, the phase-angle difference is larger between O_1 and Z (positive in the example shown) than between O_2 and Z (negative in the example shown)

sequently, Ψ is positive if the phase angle of the driven oscillator leads that of the driver and negative (as in Fig. A.2) if it lags behind it.

According to a general rule of oscillator theory Ψ increases with an increase of the ratio $\frac{T}{\tau_n}$. Hence Ψ should increase if either the zeitgeber period T increases (Fig. A.5) or the natural period τ_n decreases (Fig. A.6).

References

Armitage KB, Shulenberger E (1972) Evidence for a circannual metabolic cycle in *Citellus tridecemlineatus*, a hibernator. Comp Biochem Physiol 42A:667–688

Aschoff J (1955) Jahresperiodik der Fortpflanzung bei Warmblütern. Stud Gen 8:742–776

Aschoff J (1960) Exogenous and endogenous components in circadian rhythms. Cold Spring Harbor Symp Quant Biol 25:11–28

Aschoff J (1969) Phasenlage der Tagesperiodik in Abhängigkeit von Jahreszeit und Breitengrad. Oecologia (Berl) 3:125–165

Aschoff J (1979) Circadian rhythms: influences of internal and external factors on the period measured in constant conditions. Z Tierpsychol 49:225–249

Aschoff J (1980) Biological clocks in birds. In: Proc 17th Int Ornithol Congr Berlin, pp 113–136

Aschoff J (1981a) Free-running and entrained circadian rhythms. In: Aschoff J (ed) Handbook of behavioral neurobiology, vol 4. Plenum, New York, pp 81–93

Aschoff J (ed) (1981b) Handbook of behavioral neurobiology, vol 4. Plenum, New York

Aschoff J, Pohl H (1978) Phase relations between a circadian rhythm and its zeitgeber within the range of entrainment. Naturwissenschaften 65:80–84

Aschoff J, Klotter K, Wever R (1965) Circadian vocabulary. In: Aschoff J (ed) Circadian clocks. North-Holland, Amsterdam, pp X–XIX

Baggermann B (1957) An experimental study on the timing of breeding in the three-spined stickleback (*Gasterosteus aculeatus* L.). Arch Neerl Zool 12:105–318

Baggermann B (1980) Photoperiodic and endogenous control of the annual reproductive cycle in teleost fishes. In: Ali MA (ed) Environmental physiology of fishes. Plenum, New York, pp 533–567

Bailey SER (1981) Circannual and circadian rhythms in the snail *Helix aspera* Müller and the photoperiodic control of annual activity and reproduction. J. Comp Physiol 142:89–94

Baker JR (1938a) The relation between latitude and breeding seasons in birds. Proc Zool Soc Lond 108A:557–582

Baker JR (1938b) The evolution of breeding seasons. In: de Beer GR (ed) Evolution: essays on aspects of evolutionary biology. Oxford Univ Press, London, pp 161–177

Baker JR, Baker I (1934/36) The seasons in a tropical rain-forest (new Hebrides) II. Botany. J Linn Soc Lond Zool 39:507–519

Baker JR, Ranson RM (1938) The breeding seasons of southern hemisphere birds in the northern hemisphere. Proc Zool Soc Lond 108A:101–141

Baum MJ, Goldfoot DA (1974) Effect of hypothalamic lesions on maturation and annual cyclicity of the ferret testis. J Endocrinol 62:59–73

Beasley LJ, Pelz KM, Zucker I (1984) Circannual rhythms of body weight in pallid bats. Am J Physiol 246:R955–R958

Bellrose FC (1958) The orientation of displaced waterfowl in migration. Wilson Bull 70:20–40

Benoit J (1970) Discussion to F. Halberg: Body temperature, circadian rhythms and the eye. In: Benoit J, Assenmacher I (eds) La photorégulation de la réproduction chez les oiseaux et les mammifères. Eds Centre Natl Rech Sci Paris

Benoit J, Assenmacher I, Brard E (1955) Evolution tésticulaire du canard domestique maintenu a l'obscurité totale pendant une longue durée. CRAS Paris 241:251–253

Benoit J, Assenmacher I, Brard E (1956) Etude de l'évolution testiculaire du Canard domestique soumis très jeune à un éclairement artificiel permanent pendant deux ans. CRAS Paris 242:3113–3115

Benoit J, Assenmacher I, Brard E (1959) Action d'un éclairement permanent prolongé sur l'évolution testiculaire du Canard pekin. Arch Anat Microsc Morphol Exp 48:5–12

Benoit J, Assenmacher I, Brard E, Kordon C (1970) Influence des facteurs lumineux et sociaux sur le conditionnement testiculaire et le comportement sexuel du canard mâle de race pekin. In: Les hormones et le comportement. Probl Actuels d'Endocrinol Nutr SE 14. Expans Sci Franç, Paris 1970, pp 109–126

Berthold A (1837) Arch Anat Physiol, pp 63–68

Berthold P (1973a), Circannuale Periodik bei Teilziehern und Standvögeln. Naturwissenschaften 60:522–523

Berthold P (1973b) Relationships between migratory restlessness and migration distance in six *Sylvia* species. Ibis 115:594–599

Berthold P (1974a) Circannuale Periodik bei Grasmücken (*Sylvia*) III. Periodik der Mauser, der Nachtunruhe und des Körpergewichts bei mediterranean Arten mit unterschiedlichem Zugverhalten. J Ornithol 115:251–272

Berthold P (1974b) Circannual rhythms in birds with different migratory habits. In: Pengelley ET (ed) Circannual clocks. Academic Press, New York, pp 44–94

Berthold P (1974c) Endogene Jahresperiodik – Innere Jahreskalender als Grundlage der jahreszeitlichen Orientierung bei Tieren und Pflanzen. Konstanzer Universitätsreden 69, Univ Konstanz

Berthold P (1975a) Migratory fattening: endogenous control and interaction with migratory activity. Naturwissenschaften 62:399

Berthold P (1975b) Migration: control and metabolic physiology. In: Farner DS, King JR (eds) Avian biology, vol 5. Academic Press, New York, pp 77–128

Berthold P (1976a) Animalische und vegetabilische Ernährung omnivorer Singvogelarten: Nahrungsbevorzugung, Jahresperiodik der Nahrungswahl, physiologische und ökologische Bedeutung. J Ornithol 117:145–209

Berthold P (1976b) Über den Einfluß der Fettdeposition auf die Zugunruhe bei der Gartengrasmücke (*Sylvia borin*). Vogelwarte 28:263–266

Berthold P (1977a) Über die Entwicklung von Zugunruhe bei der Gartengrasmücke (*Sylvia borin*) bei verhinderter Fettdeposition. Vogelwarte 29:113–116

Berthold P (1977b) Über eine mögliche endogene Steuerung der Zugdisposition beim Fichtenkreuzschnabel *Loxia curvirostra*. J Ornithol 118:203–204

Berthold P (1978) Circannuale Rhythmik: freilaufende selbsterregte Periodik mit lebenslanger Wirksamkeit bei Vögeln. Naturwissenschaften 65:546

Berthold P (1979a) Über die photoperiodische Synchronisation circannualer Rhythmen bei Grasmükken., Vogelwarte 30:7–10

Berthold P (1979b) Beziehungen zwischen Zugunruhe und Zug bei der Sperbergrasmücke *Sylvia nisoria:* eine ökophysiologische Untersuchung. Vogelwarte 30:77–84

Berthold P (1980) Die endogene Steuerung der Jahresperiodik: eine kurze Übersicht. Proc 17th Int Ornithol Congr Berlin, pp 473–478

Berthold P, Leisler B (1980) Migratory restlessness of the marsh warbler, *Acrocephalus palustris*. Naturwissenschaften 67:472

Berthold P, Querner U (1981) Genetic basis of migratory behavior in European warblers. Science 212:77–79

Berthold P, Querner U (1982) Genetic basis of molt, wing length, and body weight in a migratory bird species, *Sylvia atricapilla*. Experientia (Basel) 38:801–802

Berthold P, Gwinner E, Klein H (1970) Vergleichende Untersuchung der Jugendentwicklung eines ausgeprägten Zugvogels, *Sylvia borin*, und eines weniger ausgeprägten Zugvogels, *S. atricapilla*. Vogelwarte 25:297–331

Berthold P, Gwinner E, Klein H (1971) Circannuale Periodik bei Grasmücken (*Sylvia*). Experientia (Basel) 27:399

Berthold P, Gwinner E, Klein H (1972a) Circannuale Periodik bei Grasmücken I. Periodik des Körpergewichts, der Mauser und der Nachtunruhe bei *Sylvia atricapilla* und *S. borin* unter verschiedenen konstanten Bedingungen. J Ornithol 113:170–190

Berthold P, Gwinner E, Klein H (1972b) Circannuale Periodik bei Grasmücken II. Periodik der Gonadengröße bei *Sylvia atricapilla* und *S. borin* unter verschiedenen, konstanten Bedingungen. J Ornithol 113:407–417

Berthold P, Gwinner E, Querner U (1974) Vergleichende Untersuchung der Jugendentwicklung südfinnischer und südwestdeutscher Gartengrasmücken, *Sylvia borin*. Ornis Fenn 51:146–154

Bhatt A, Chandola A (1985) Circannual rhythm of food intake in spotted munia and its phase relationship with fattening and reproductive cycles. J Comp Physiol A Sens Neural Behav Physiol 156:429–432

Biebach H (1985) Sahara stopover in migratory flycatchers: fat and food affect the time program. Experientia (Basel) 41:695–697

Binkley S (1977) Constant light: effects on the circadian locomotor rhythm in the house sparrow. Physiol Zool 50:170–181

Bittman EL, Karsch FJ, Hopkins JW (1983a) Role of the pineal gland in ovine photoperiodism: regulation of seasonal breeding and negative feedback effects of estradiol upon luteinizing hormone secretion. Endocrinology 113:329–336

Bittman EL, Dempsey RJ, Karsch FJ (1983b) Pineal melatonin secretion drives the reproductive response to daylength in the ewe. Endocrinology 113:2276–2283

Blake GM (1959a) Control of diapause by an "internal clock" in *Anthrenus verbasci* (L.) (*Col. Dermestidae*). Nature 183:126–127

Blake GM (1959b) Diapause and the regulations of development in *Anthrenus verbasci* (L.). (*Col. Dermestidae*). Bull Entomol Res 49:751–757

Blake GM (1960) Decreasing photoperiod inhibiting metamorphosis in an insect. Nature 188:168–169

Blake GM (1963) Shortening of a diapause-controlled life cycle by means of increasing photoperiod. Nature 198:462–463

Blanchard BD (1941) The white-crowned sparrow (*Zonotrichia leucophrys*) of the pacific seaboard: environment and annual cycle. Univ Calif Publ Zool 46:1–178

Blanchard BD, Erickson MM (1949) The cycle in the Gambel sparrow. Univ Calif Publ Zool 47:255–318

Bluymental TI, Dolnik VR (1966) Geographical and intrapopulational differentiation in the time of breeding, molt and migration in some migratory passerines. Izd Ural Filiala Acad Nauk SSSR, Sverdlovsk, pp 319–332

Bornkamm (1966) Ein Jahresrhythmus des Wachstums bei *Lemna minor* L. Planta (Berl) 69:178–186

Brehm CL (1828) Der Zug der Vögel. Ibis 21:912–922

Brock MA (1975a) Circannual rhythms I. Free-running rhythms in growth and development of the marine cnidarian, *Campanularia flexuosa*. Comp Biochem Physiol 51A:377–383

Brock MA (1975b) Circannual rhythms II. Temperature-compensated free-running rhythms in growth and development of the marine cnidarian, *Campanularia flexuosa*. Comp Biochem Physiol 51A:385–390

Brock MA (1975c) Circannual rhythms III. Rhythmicity in the longevity of hydranths of the marine cnidarian, *Campanularia flexuosa*. Comp Biochem Physiol 51A:391–398

Brown ME (1945) The growth of brown trout (*Salmo trutta* L.) II. The growth of two-year-old trout at a constant temperature of 11.5 °C. J Exp Biol 22:130–144

Bünning E (1936) Die endogene Tagesperiodik als Grundlage der photoperiodischen Reaktion. Ber Dtsch Bot Ges 54:590–608

Bünning E (1948) Zur Kenntnis der endogenen Jahresrhythmik in Samen. Naturwissenschaften 35:221–222

Bünning E (1949a) Jahres- und tagesperiodische Vorgänge in der Pflanze. Stud Gen 2:73–78

Bünning E (1949b) Zur Physiologie der endogenen Jahresrhythmik in Pflanzen, speziell in Samen. Z Naturforsch Teil B 4:167–176

Bünning E (1955) Jahreszeiten und Pflanzenleben. Stud Gen 8:733–742

Bünning E (1956) Endogene Aktivitätsrhythmen. Encycl Plant Physiol 2:878–907

Bünning E (1973) The physiological clock (3rd ed). Springer, Berlin Heidelberg New York

Bünning E, Bauer EW (1952) Über die Ursachen endogener Keimfähigkeitsschwankungen im Samen. Z Bot 40:67–76

Bünning E, Müssle L (1951) Der Verlauf der endogenen Jahresrhythmik in Samen unter dem Einfluß verschiedenartiger Außenfaktoren. Z Naturforsch Teil B 6:108–112

Butschke HW (1977) Investigations of circadian and circannian rhythms in field-caught and "Kaspar-Hauser" fat dormice (*Glis glis* L.) with self-selection experiments. J Interdiscip Cycle Res 8:293–296

Canguilhem B (1985) Rythmes circannuels chez les mammifères hibernants sauvages. Can J Zool 63:453–464

Canguilhem B, Bloch R (1967) Evolution saisonnière de l'élimination des hormones surrénaliennes chez un hibernant, *Cricetus cricetus*. Arch Sci Physiol 21:27–43

Canguilhem B, Schieber JP, Koch A (1973) Rythme circannuel ponderal du hamster d'Europe (*Cricetus cricetus*). Influences respectives de la photopériode et de la température externe sur son deroulement. Arch Sci Physiol 27:67–90

Canguilhem B, Schmitt P, Mack P, Kempf E (1977) Comportement alimentaire, rythmes circannuels ponderals et d'hibernation chez le hamster d'Europe porteur de lesions des fasceaux noradrenergiques ascendants. Physiol Behav 18:1067–1074

Chandola A, Pathak VK, Bhatt D (1982) Evidence for an endogenous circannual component in the control of the annual gonadal cycle in spotted munia. J Interdicip Cycle Res 13:281–286

Chapin JP (1932) The birds of the Belgian Congo, vol 1. Bull Am Mus Nat Hist 65:1–756

Chapin JP (1954) The calender of wideawake fair. Auk 71:1-15

Chapin JP, Wing LW (1959) The wideawake calendar, 1953–1958. Auk 76:153–158

Craig AJFK (1985) Breeding condition of male red bishops under artificial photoperiods. Ostrich 56:74–78

Cranford JA (1978) Hibernation in the western jumping mouse (*Zapus princeps*). J Mammal 59:496–509

Curry-Lindahl K (1963) Molt, body weights, gonadal development, and migration in *Motacilla flava*. Proc 18th Int Ornithol Congr Ithaca, pp 960–973

Czeschlik D (1976) Der Einfluß des Wetters auf die Zugunruhe von Garten- und Mönchsgrasmücken (*Sylvia borin* und *S. atricapilla*). Ph. D. Thesis, Univ Innsbuck

Daan S (1973) Periodicity of heterothermy in the garden dormouse, *Eliomys quercinus* (L.). Neth J Zool 23:237–265

Daan S, Aschoff J (1975) Circadian rhythms of locomotor activity in captive birds and mammals: their variations with season and latitude. Oecologia (Berl) 18:269–316

Danilevskii AS (1965) Photoperiodism and seasonal development of insects. Oliver and Boyd, Edinburgh

Dark J, Zucker I (1986) Circannual rhythms of ground squirrels: role of the hypothalamic paraventricular nucleus. J Biol Rhythms 1:17–23

Dark J, Pickard GE, Zucker I (1985) Persistence of circannual rhythms in ground squirrels with lesions of the suprachiasmatic nuclei. Brain Res 332:201–207

Davis DE (1967) The annual rhythm of fat deposition in woodchucks (*Marmota monax*). Physiol Zool 40:391–402

Davis DE (1976) Hibernation and circannual rhythms of food consumption in marmots and ground squirrels. Q Rev Biol 51:477–514

Davis DE (1984) Circannual changes in nesting behavior of captive ground squirrels (*Spermophilus beecheyi*). J Interdiscip Cycle Res 14:189–194

Davis DE, Finnie EP (1975) Entrainment of circannual rhythm in weight of woodchucks. J Mammal 56:199–203

Davis DE, Swade RH (1983) Circannual rhythm of torpor and molt in the ground squirrel, *Spermophilus beecheyi*. Comp Biochem Physiol 76A:183–187

Dawson A, Goldsmith AR (1982) Prolactin and gonadotrophin secretion in wild starlings (*Sturnus vulgaris*) during the annual cycle and in relation to nesting, incubation, and rearing young. Gen Comp Endocrinol 48:213–221

Dawson A, Goldsmith AR (1983) Plasma prolactin and gonadotrophins during gonadal development and the onset of photorefractoriness in male and female starlings (*Sturnus vulgaris*) on artificial photoperiods. J Endocrinol 97:253–260

Dementiew GP, Gladkow NA (1954) The birds of the Soviet Union, vol 6. Sovetskaya Nauka, Moscow

Dingler H (1911) Versuche über die Periodizität einiger Holzgewächse in den Tropen. Sitzungsber Bayer Akad Wiss München, Math-Phys Klasse, p 127

Dolnik VR (1974) Annual cycles of migratory fat deposition, sexual activity, and molt in chaffinches (*Fringilla coelebs*) under constant photoperiodic conditions. Z Observ Biol 35:543–555

Dolnik VR (1976) In: Zaslavsky VA (ed) Fotoperiodizm Zhivotnykh i Rastenii. Akad Nauk SSSR, Leningrad, pp 47 ff

Donham RS, Moore MC, Farner DS (1983) Physiological basis of repeated testicular cycles on 12-hour days (12L 12D) in white-crowned sparrows, *Zonotrichia leucophrys gambelii*. Physiol Zool 56:302–307

Dorka V (1966) Das jahres- und tageszeitliche Zugmuster von Kurz- und Langstreckenziehern nach Beobachtungen auf den Alpenpässen Cou/Bretolet (Wallis). Ornithol Beob 63:165–223

Drost R (1938) Über den Einfluß von Verfrachtungen zur Herbstzugzeit auf den Sperber, *Accipiter nisus* (L.). Zugleich ein Beitrag zur Frage nach der Orientierung der Vögel auf dem Zuge ins Winterquartier. 9th Int Ornithol Congr Rouen, pp 503–521

Dubois R (1896) Étude sur le méchanisme de la thermogenèse et du someil chez les mammifères. Physiologie comparée de la marmotte. Ann Univ Lyon 25:1–268

Ducker MJ, Bowman JC, Temple A (1973) The effect of constant photoperiod on the expression of oestrus in the ewe. J Reprod Fertil (Suppl) 19:143–150

Elliott JA (1976) Circadian rhythms and photoperiodic time measurement in mammals. Fed Proc 35:2339–2346

Engelmann W (1986) Effects of lithium salts on circadian rhythms. In: Halaris, A (ed) Neuropsychiatric disorders and disturbances in the circadian system of man (in press)

Enright JT (1970) Ecological aspects of endogenous rhythmicity. Annu Rev Ecol Syst 1:221–238

Enright JT (1971) Heavy water slows biological timing processes. Z Vergl Physiol 72:1–16

Eriksson LU, Lundquist H (1982) Circannual rhythms and photoperiod regulation of growth and smolting in baltic salmon (*Salmo salar* L.). Aquaculture 28:113–121

Erkinaro E (1972) Seasonal changes in the phase position of circadian activity rhythms in some voles and their endogenous component. Aquilo Ser Zool 13:87–91

Eskes GA, Zucker I (1978) Photoperiodic control of hamster testis: dependence on circadian rhythms. Proc Natl Acad Sci USA 75:1034–1038

Falk H, Gwinner E (1983) Photoperiodic control of testicular regression in the European starling. Naturwissenschaften 70:257–258

Fall MW (1971) Seasonal variations in the food consumption of woodchucks (*Marmota monax*). J Mammal 52:370–375

Farner DS (1970) Predictive functions in the control of annual cycles. Environ Res 3:119–131

Farner DS (1975) Photoperiodic controls in the secretion of gonadotrophins in birds. Am Zool 15 (Suppl 1):117–135

Farner DS (1985) Annual rhythms. Ann Rev Physiol 47:65–82

Farner DS, Follett BK (1979) Reproductive periodicity in birds. In: Barrington EJW (ed) Hormones and evolution. Academic Press, London, pp 129–148

Farner DS, Gwinner E (1980) Photoperiodicity, circannual and reproductive cycles. In: Epple A, Stetson MH (eds) Avian endocrinology. Academic Press, New York, pp 331–366

Farner DS, Lewis RA (1971) Photoperiodism and reproductive cycles in birds. In: Giese AC (ed) Photophysiology, vol 6. Academic Press, New York, pp 325–370

Farner DS, Serventy DL (1960) The timing of reproduction in birds in the arid regions of Australia. Anat Res 137:354

Farner DS, Wingfield JG (1980) Reproductive endocrinology of birds. Ann Rev Physiol 42:455–470

Farner DS, Lewis RA, Darden TR (1973) Photoperiodic control mechanisms. In: Altman PL, Dittmer DS (eds) Biological data book (2nd ed), vol 2. Fed Am Soc Exp Biol, pp 1047–1052

Farner DS, Donham RS, Moore MC, Lewis RA (1980) The temporal relationship between the cycle of testicular development and molt in the white-crowned sparrow, *Zonotrichia leucophrys gambelii*. Auk 97:63–75

Farner DS, Donham RS, Matt KS, Mattocks PW, Moore MC, Wingfield JC (1983) The nature of photorefractoriness. In: Mikami S, Homna K, Wada M (eds) Avian endocrinology: environmental and ecological perspectives. Japan Sci Soc Press, Tokyo. Springer, Berlin Heidelberg New York, pp 149–166

Fischer K, Butschke HW, Mahlert D (1975) Untersuchungen zur circadianen und circannualen Rhythmik bei Siebenschläfern (*Glis glis* L.) im Selbstwählversuch. Z Säugetierk 40:65–74

Fogden MP (1972) The seasonality and population dynamics of equatorial forest birds in Sarawak. Ibis 114:307–343

Follett BK (1973) Circadian rhythms and photoperiodic time measurement in birds. J Reprod Fertil (Suppl) 19:5–18

Follett BK, Davies DT (1975) Photoperiodicity and the neuroendocrine control of reproduction in birds. Symp Zool Soc Lond 35:199–224

Follett BK, Robinson JE (1980) Photoperiod and gonadotrophin secretion in birds. Progr Reprod Biol, vol 5. Karger, Basel, pp 39–61

French AR (1977) Circannual rhythmicity and entrainment of surface activity in the hibernator *Perognathus longimembris*. J Mammal 58:37–43

Fuller CA, Lydic R, Sulzman FM, Albers HE, Tepper B, Moore-Ede MC (1981) Circadian rhythm in body temperature persists after suprachiasmatic lesions in the squirrel monkey. Am J Physiol 241:R385–R391

Funakoshi K, Uchida TA (1982) Annual cycles of body weight in the Nannie's frosted bat, *Vespertilio superans superans*. J Zool (Lond) 196:417–430

Gänshirt G, Gwinner E (1979) Jahresperiodik der Gonadengröße und der Mauser beim Star (*Sturnus vulgaris*) unter Photoperiodezyklen unterschiedlicher Amplitude. J Ornithol 120:322–325

Gänshirt G, Daan S, Gerkema MP (1984) Arrhythmic perch hopping and rhythmic feeding of starlings in constant light: separate circadian oscillators? Comp Physiol A Sens Neural Behav Physiol 154:669–674

Goldman BD, Darrow JM (1983) The pineal gland in mammalian photoperiodism. Neuroendocrinology 37:386–396

Goldsmith AR, Nicholls TJ (1984a) Thyroxine induces photorefractoriness and stimulates prolactin secretion in European starlings (*Sturnus vulgaris*). J Endocrinol 101:R1–R3

Goldsmith AR, Nicholls TJ (1984b) Prolactin is associated with the development of photorefractoriness in starlings. Gen Comp Endocrinol 54:247–255

Goldsmith AR, Nicholls TJ (1984c) Thyroidectomy prevents the development of photorefractoriness and the associated rise in plasma prolactin in starlings. Gen Comp Endocrinol 54:256–263

Goss RJ (1969a) Photoperiodic control of antler cycles in deer I. Phase shift and frequency changes. J Exp Zool 170:311–324

Goss RJ (1969b) Photoperiodic control of antler cycles in deer II. Alterations in amplitude. J Exp Zool 171:223–234

Goss RJ (1976) Photoperiodic control of antler cycles in deer III. Decreasing versus increasing daylengths. J Exp Zool 197:307–312

Goss RJ (1980) Photoperiodic control of antler cycles in deer V. Reversed seasons. J Exp Zool 211:101–105

Goss RJ (1984) Photoperiodic control of antler cycles in deer VI. Circannual rhythms on altered day lenghts. J Exp Zool 230:265–271

Gwinner E (1966) Tagesperiodische Schwankungen der Vorzugshelligkeit bei Vögeln. Z Vergl Physiol 52:370–379

Gwinner E (1967) Circannuale Periodik der Mauser und der Zugunruhe bei einem Vogel. Naturwissenschaften 54:447

Gwinner E (1968a) Circannuale Periodik als Grundlage des jahreszeitlichen Funktionswandels bei Zugvögeln. Untersuchungen am Fitis (*Phylloscopus trochilus*) und am Waldlaubsänger (*P. sibilatrix*). J Ornithol 109:70–95

Gwinner E (1968b) Artspezifische Muster der Zugunruhe bei Laubsängern und ihre mögliche Bedeutung für die Beendigung des Zuges im Winterquartier. Z Tierpsychol 25:843–853

Gwinner E (1971a) A comparative study of circannual rhythms in warblers. In: Menaker M (ed) Biochronometry. Natl Acad Sci, Wash DC, pp 405–427

Gwinner E (1971b) Orientierung. In: Schüz E (ed) Grundriß der Vogelzugskunde. Parey, Berlin, pp 299–348

Gwinner E (1972a) Endogenous timing factors in bird migration. In: Galler SR et al. (eds) Animal orientation and navigation. Natl Acad Sci, Wash DC, pp 321–338

Gwinner E (1972b) Adaptive functions of circannual rhythms in warblers. Proc 15th Int Ornithol Congr The Hague, pp 218–236

Gwinner E (1973) Circannual rhythms in birds: their interaction with circadian rhythms and environmental photoperiod. J Reprod Fertil (Suppl) 19:51–65

Gwinner E (1974a) Testosterone induces "splitting" of circadian locomotor activity rhythms in birds. Science 185:72–74

Gwinner E (1974b) Endogenous temporal control of migratory restlessness in warblers. Naturwissenschaften 61:405

Gwinner E (1975a) Die circannuale Periodik der Fortpflanzungsaktivität beim Star (*Sturnus vulgaris*) unter dem Einfluß gleich- und andersgeschlechtiger Artgenossen. Z Tierpsychol 38:34–43

Gwinner E (1975b) Effects of season and external testosterone on the free-running circadian activity rhythm of European starlings (*Sturnus vulgaris*). J Comp Physiol 103:315–328

Gwinner E (1975c) Circadian and circannual rhythms in birds. In: Farner DS, King JR (eds) Avian biology. vol 5. Academic Press, New York, pp 221–285

Gwinner E (1977a) Über die Synchronisation circannualer Rhythmen bei Vögeln. Vogelwarte 29:16–25

Gwinner E (1977b) Circannual rhythms in bird migration. Annu Rev Ecol Syst 8:381–405

Gwinner E (1978) Effects of pinealectomy on circadian locomotor activity rhythms in European starlings (*Sturnus vulgaris*). J Comp Physiol 126:123–129

Gwinner E (1979) Jugendentwicklung südfinnischer und süddeutscher Gartengrasmücken (*Sylvia borin*) unter denselben Bedingungen. Vogelwarte 30:41–43

Gwinner E (1980) Relationship between circadian activity parameters and gonadal function: evidence for internal coincidence? Proc 17th Int Ornithol Congr Berlin 1980, pp 409–416

Gwinner E (1981a) Circannual rhythms: their dependence on the circadian system. In: Follett BK, Follett DE (eds) Biological clocks in seasonal reproductive cycles. Wright, Bristol, pp 153–169

Gwinner E (1981b) Circannuale Rhythmen bei Tieren und ihre photoperiodische Synchronisation. Naturwissenschaften 68:542–551

Gwinner E (1981c) Annual rhythms: perspective. In: Aschoff J (ed) Handbook of behavioral neurobiology, vol 4. Plenum, New York, pp 381–389

Gwinner E (1981d) Circannual systems. In: Aschoff J (ed) Handbook of behavioral neurobiology, vol 4. Plenum, New York, pp 391–410

Gwinner E (1983) Änderungen der Zugunruhe, des Körpergewichts und der Mauser von Dorngrasmücken (*Sylvia communis*) unter einer konstanten 12-stündigen Photoperiode. Vogelwarte 32:77–80

Gwinner E (1986) Circannual rhythms in the control of avian migrations. Adv Stud Behav 16:191–228

Gwinner E, Biebach H (1977) Endogene Kontrolle der Mauser und der Zugdisposition bei südfinnischen und südfranzösischen Neuntötern (*Lanius collurio*). Vogelwarte 29:56–63

Gwinner E, Dittami J (1980) Pinealectomy affects the circannual testicular rhythm in European starlings (*Sturnus vulgaris*). J Comp Physiol 136:345–348

Gwinner E, Dittami J (1982) Pineal influences on circannual cycles in European starlings: effects through the circadian system? In: Aschoff J (ed) Vertebrate circadian systems: structure and physiology. Springer, Berlin Heidelberg New York, pp 271–284

Gwinner E, Dittami J (1985) Photoperiodic responses in temperate zone and equatorial stonechats: a contribution to the problem of photoperiodism in tropical organisms. In: Follett BK, Ishii S, Chandola A (eds) The endocrine system and the environment. Japan Sci Soc Press, Tokyo. Springer, Berlin Heidelberg New York, pp 279–294

Gwinner E, Dittami J (1986) Adaptive functions of circannual clocks. In: Boissin J (ed) Endocrine regulations as adaptive mechanisms to the environment. CNRS Publ (in press)

Gwinner E, Dorka V (1976) Endogenous control of annual biological rhythms in birds. Proc 16th Int Ornithol Congr Canberra, pp 223–234

Gwinner E, Eriksson LO (1977) Circadiane Rhythmik und photoperiodische Zeitmessung beim Star (*Sturnus vulgaris*). J Ornithol 118:60–67

Gwinner E, Schwabl-Benzinger I (1982) Adaptive temporal programming of molt and migratory disposition in two closely related long-distance migrants, the pied flycatcher (*Ficedula hypoleuca*) and the collared flycatcher (*F. albicollis*). In: Papi F, Wallraff H (eds) Avian navigation. Springer, Berlin Heidelberg New York, pp 75–89

Gwinner E, Wiltschko W (1978) Endogenously controlled changes in migratory direction of the garden warbler, *Sylvia borin*. J Comp Physiol 125:267–273

Gwinner E, Wiltschko W (1980) Circannual changes in migratory orientation of the garden warbler, *Sylvia borin*. Behav Ecol Sociobiol 7:73–78

Gwinner E, Wozniak J (1982) Circannual rhythms in European starlings: Why do they stop in long photoperiods? J Comp Physiol 146:419–421

Gwinner E, Berthold P, Klein H (1971) Untersuchungen zur Jahresperiodik von Laubsängern II. Einfluß der Tageslichtdauer auf die Entwicklung des Gefieders, des Gewichts und der Zugunruhe bei *Phylloscopus trochilus* und *Ph. collybita*. J Ornithol 112:253–265

Gwinner E, Berthold P, Klein H (1972) Untersuchungen zur Jahresperiodik von Laubsängern III. Die Entwicklung des Gefieders, des Gewichts und der Zugunruhe südwestdeutscher und skandinavischer Fitisse (*Phylloscopus trochilus trochilus* and *Ph. t. acredula*). J Ornithol 113:1–8

Gwinner E, Dittami J, Gänshirt G (1980) Gibt es im circannualen Hodenzyklus des Stars (*Sturnus vulgaris*) eine Refraktärphase? Vogelwarte 30:335–337

Gwinner E, Wozniak J, Dittami J (1981) The role of the pineal organ in the control of annual rhythms. In: Oksche A, Pevet P (eds) The pineal organ: photobiology, biochronometry, endocrinology. Elsevier North-Holland, pp 99–121

Gwinner E, Biebach H, von Kries I (1985a) Food availability affects migratory restlessness in caged garden warblers (*Sylvia borin*). Naturwissenschaften 72:51–52

Gwinner E, Dittami J, Gänshirt G, Hall M, Wozniak J (1985b) Endogenous and exogenous components in the control of the annual reproductive cycle of the European starling. Proc 18th Int Ornithol Congr Moscow, 501–515

Haberey P, Canguilhem B, Kayser C (1967) Evolution saisonnière de l'élimination urinaire du sodium et du potassium chez le Hamster d'Europa (Cricetus cricetus). Comptes Rendus des Séances de la Société de biologie et de ses filiales 161:2044–2048

Hall VD, Goldman BD (1982) Hibernation in the female Turkish hamster (*Mesocricetus brandti*): An investigation of the role of the ovaries and of photoperiod. Biol Reprod 27:811–815

Hamner WM (1971) On seeking an alternative to the endogenous reproductive rhythm hypothesis in birds. In: Menaker M (ed) Biochronometry. Natl Acad Sci, Wash DC, pp 428–447

Hamner WM, Enright JT (1967) Relationships between photoperiodism and circadian rhythms of activity in the house finch. J Exp Biol 46:43–61

Hamner WH, Stocking J (1970) Why don't bobolinks breed in Brazil? Ecology 51:743–751

Hastings MH, Herbert J, Martensz ND, Roberts AC (1985) Annual reproductive rhythms in mammals: mechanisms of light synchronization. Ann NY Acad Sci 453:182–204

Heller HC, Poulson TL (1970) Circannian rhythms II. Endogenous and exogenous factors controlling reproduction and hibernation in chipmunks (*Eutamias*) and ground squirrels (*Spermophilus*). Comp Biochem Physiol 33:357–383

Hengst RA, Wiebers JE (1984) A circannual cycle in the core-to-skin temperature gradient of normothermic thirteen-lined ground squirrels. J Interdiscip Cycle Res 15:315–320

Henssen A (1954) Die Dauerorgane von *Spirodela polyrrhiza* (L.) Schleid, in physiologischer Betrachtung. Flora (Jena) 141:523–566

Herbert J (1971) The role of the pineal gland in the control by light of the reproductive cycle of the ferret. In: Wolstenholme GEW, Knight J (eds) The pineal gland. Ciba Symp. Churchill, London, pp 303–327

Herbert J (1972) Initial observations on pinealectomized ferrets kept for long periods in either daylight or artificial illumination. J Endocrinol 55:591–597

Herbert J, Stacey PM, Thorpe DH (1978) Recurrent breeding seasons in pinealectomized or optic-nerve-sectioned ferrets. J Endocrinol 78:389–397

Hoar WS (1969) Reproduction. In: Hoar WS, Randall DJ (eds) Fish physiology. Academic Press, New York, pp 1–72

Hock RJ (1955) Photoperiod as a stimulus for onset of hibernation. Fed Proc 14:73–74

Hock RJ (1956) Body temperature variations of non-hibernating Alaskan ground squirrels. Fed Proc 16:440

Hoffmann K (1969) Die relative Wirksamkeit von Zeitgebern. Oecologia (Berl) 3:184–206

Hoffmann K (1981) Photoperiodism in vertebrates. In: Aschoff J (ed) Handbook of behavioral neurobiology, vol 4. Plenum, New York, pp 449–473

Holmes RT (1966) Molt cycle of the red-backed sandpiper (*Calidris alpina*) in western North America. Auk 83:517–533

Homeyer EF von (1881) Die Wanderungen der Vögel. Grieben's, Leipzig

Howles CM, Webster GM, Haynes NB (1980) The effect of rearing under a long or short photoperiod on testis growth, plasma testosterone and prolactin concentrations, and the development of sexual behavior in rams. J Reprod Fertil 60:437–447

Howles CM, Craigon J, Haynes NB (1982) Long-term rhythms of testicular volume and plasma prolactin concentrations in rams reared for 3 years in constant photoperiod. J Reprod Fertil 65:439–446

Hutton KE (1960) Seasonal physiological changes in the red-eared turtle, *Pseudemys scripta elegans*. Copeia 4:360–362

Ihering H von (1923) Der periodische Blattwechsel der Bäume im tropischen und subtropischen Südamerika. Englers Jahrb Bot 58:524–598

Immelmann K (1963a) Tierische Jahresperiodik in ökologischer Sicht. Zool Jahrb Abt Syst Oekol Geogr Tiere 91:91–200

Immelmann K (1963b) Drought adaptations in Australian desert birds. Proc 18th Int Ornithol Congr Ithaca, pp 649–657

Immelmann K (1967) Periodische Vorgänge in der Fortpflanzung tierischer Organismen. Stud Gen 20:15–33

Immelmann K (1971) Ecological aspects of periodic reproduction. In: Farner DS, King JR (eds) Avian biology, vol 1. Academic Press, New York, pp 341–389

Immelmann K (1973) Role of the environment in reproduction as source of "predictive" information. In: Farner DS (ed) Breeding biology of birds. Natl Acad Sci, Wash DC, pp 121–147

Jallageas M, Assenmacher I (1984) External factors controlling annual testosterone and thyroxine cycles in the edible dormouse *Glis glis*. Comp Biochem Physiol 77A:161–167

Jander R (1963) Insect orientation. Annu Rev Entomol 8:95–114

Jegla TC, Poulson TL (1970) Circannian rhythms I. Reproduction in the cave crayfish, *Orconectes pellucidus inermis*. Comp Biochem Physiol 33:347–355

Joy JE (1984) Population differences in circannual cycles of thirteen-lined ground squirrels. In: Michener GR, Murie JD (eds) Biology of ground-dwelling sciurids. Univ Nebraska Press, Lincoln, pp 125–141

Joy JE, Mrosovsky N (1982) Circannual cycles of molt in ground squirrels. Can J Zool 60:3227–3231

Joy JE, Mrosovsky N (1983) Circannual cycles in golden-mantled ground squirrels: lengthening of period by low temperatures in the spring phase. J Comp Physiol 150:23–238

Joy JE, Mrosovsky N (1985) Synchronization of circannual cycles: a cold spring delays the cycles of thirteen-lined ground squirrels. J Comp Physiol A Sens Neural Behav Physiol 156:125–134

Joy JE, Melnyk RB, Mrosovsky N (1980) Reproductive cycles in the male dormouse (*Glis glis*). Comp Biochem Physiol 67A:219–221

Kavaliers M (1982) Seasonal and circannual rhythms in behavioral thermoregulation and their modifications by pinealectomy in the white sucker, *Catostomus commersoni*. J Comp Physiol 146:235–243

Kayser C (1940) Essai d'analyse du méchanisme du sommeil hibernal. Ann Physiol Veg (Paris) 16:313–372

Keast A (1959) Australian birds: their zoogeography and adaptations to an arid environment. In: Keast A, Crocher RL, Christian CS (eds) Biogeography and ecology in Australia. Junk, The Hague, pp 89–114

Keast A (1968) Moult in birds of the Australian dry country relative to rainfall and breeding. J Zool (Lond) 155:185–200

Keast A, Marshall AJ (1954) The influence of drought and rainfall on reproduction in Australian desert birds. Proc Zool Soc Lond 124:493–499

Kempf E, Mack G, Canguilhem B, Mandel P (1978) Seasonal changes in the levels and the turnover of brain serotonin and noradrenaline in the European hamster kept under constant environment. Experientia 34:1032–1033

Kenagy GJ (1980) Interrelation of annual rhythms of reproduction and hibernation in the golden-mantled ground squirrel. J Comp Physiol 135:333–339

Kenagy GJ (1981a) Effects of day length, temperature, and endogenous control on annual rhythms of reproduction and hibernation in chipmunks (*Eutamias* spp.) J Comp Physiol 141:369–378

Kenagy GJ (1981b) Endogenous annual rhythm of reproductive function in the non-hibernating desert ground squirrel, *Ammospermophilus leucurus*. J Comp Physiol 142:251–258

Kennaway DJ, Sanford LM, Godfrey B, Friesen HG (1983) Patterns of progesterone, melatonin, and prolactin secretion in ewes maintained in four different photoperiods. J Endocrinol 97:229–242

Kessler E, Cygan CF (1963) Seasonal changes in the nitrate-reducing activity of a green alga. Experientia (Basel) 19:89–90

Kiepenheuer J (1984) The magnetic compass mechanism of birds and its possible association with the shifting course directions of migrants. Behav Ecol Sociobiol 14:81–99

King J (1968) Cycles of fat deposition and molt in white-crowned sparrows in constant environmental conditions. Comp Biochem Physiol 24:827–837

Klein H (1972) The adaptational value of internal annual clocks in birds. In Pengelley ET (ed) Circannual Clocks. Academic Press, New York, pp 347–391

Koriba KV (1948) On the origin and meaning of deciduousness viewed from the seasonal habit of trees in the tropics I and II. Seiri seitai (Jap)

Kreuels T, Joerres R, Martin W, Brinkmann K (1984) System analysis of the circadian rhythm of *Euglena gracilis*, II: Masking effects and mutual interactions of light and temperature responses. Z Naturforsch Sect C 39:801–811

Kummerow J (1963) Endogenous fluctuations of germination capacity in *Dactylis glomerata*. J Bot 50:915–920

Lack D (1950) The breeding seasons of European birds. Ibis 92:288–316

Lapeyronie MA (1968) Existence d'un cycle endogène concernant la faculté germinative de l'*Oryzopsis miliacea*. CRAS Paris 267:1724–1726

Legan SJ, Karsch FJ (1979) Neuroendocrine regulation of the estrous cycle and seasonal breeding in the ewe. Biol Reprod 20:74–85

Legan SJ, Karsch FJ (1983) Importance of retinal photoreceptors to the photoperiodic control of seasonal breeding in the ewe. Biol Reprod 29:316–325

Legan SJ, Winans SS (1981) The photoneuroendocrine control of seasonal breeding in the ewe. Gen Comp Endocrinol 45:317–328

Licht P, Zucker I, Hubbard G, Boshes M (1982) Circannual rhythms of plasma testosterone and luteinizing hormone levels in golden-mantled ground squirrels (*Spermophilus lateralis*). Biol Reprod 27:411–418

Lincoln GA (1979) Photoperiodic control of seasonal breeding in the ram: participation of the cranial sympathetic nervous system. J Endocrinol 82:135–147

Lincoln GA (1984) Central effects of photoperiod on reproduction in the ram revealed by the use of a testosterone clamp. J Endocrinol 103:233–241

Lincoln GA, Short RV (1980) Seasonal breeding: nature's contraceptive. Recent Prog Horm Res 36:1–52

Lindsay DR, Pelletier J, Pisselet C, Conrot M (1984) Changes in photoperiod and nutrition and their effect on testicular growth of rams. J Reprod Fertil 71:351–356

Linzell JL (1973) Innate seasonal oscillations in the rate of milk secretion in goats. J Physiol (Lond) 230:225–233

Lofts B (1962) Photoperiod and the refractory period of reproduction in an equatorial bird, *Quelea quelea*. Ibis 104:407–414

Lofts B (1964) Evidence of an autonomous reproductive rhythm in an equatorial bird (*Quelea quelea*). Nature 201:523–524

Lofts B (1974) Reproduction. In: Lofts B (ed) Physiology of the amphibia, vol 2. Academic Press, New York, pp 107–218

Lyman CP (1948) The oxygen consumption and temperature regulation of hibernating hamsters. J Exp Zool 109:55–78

Lyman CP (1954) Activity, food consumption, and hoarding in hibernators. J Mammal 35:545–552

Marshall AJ (1951) The refractory period of testis rhythm in birds and its possible bearing on breeding and migration. Wilson Bull 63:238–261

Marshall AJ (1959) Internal and environmental control of breeding. Ibis 101:456–478

Marshall AJ (1960a) The role of the internal rhythm of reproduction in the timing of avian breeding seasons including migration. Proc 12th Int Ornithol Congr Helsinki, pp 475–482

Marshall AJ (1960b) Annual periodicity in the migration and reproduction of birds. Cold Spring Harbor Symp Quant Biol 25:499–505

Marshall AJ, Serventy DL (1959) Experimental demonstration of an internal rhythm of reproduction in a trans-equatorial migrant, the short-tailed shearwater *Puffinus tenuirostris*. Nature 184:1704–1705

Matt KS (1982) Seasonal regulation of gonadotropin secretion by androgen feedback in the male white-crowned sparrow (*Zonotrichia leucophrys gambelii*). Ph. D Diss, Univ Washington

Mauleon P, Rougeot J (1962) Régulation des saisons sexuelles chez des brebis de race différentes au moyen de divers rhythmes lumineux. Ann Biol Anim Biochem Biophys 2:209–222

Mayersbach H von (1978) Die Zeitstruktur des Organismus. Arzneim Forsch 28(II):1824–1836

Meier AH, Wilson JM (1985) Resetting annual cycles with neurotransmitter-affecting drugs. In: Follett BK, Ishii S, Chandola A (eds) The endocrine system and the environment. Japan Sci Soc Press, Tokyo. Springer, Berlin Heidelberg New York, pp 149–156

Meier AH, Ferrell BR, Miller LJ (1980) Circadian components of the circannual mechanisms in the white-throated sparrow. Proc 17th Int Ornithol Congr Berlin, pp 458–462

Melnyk RB (1979) Persistence of body weight cycles in dormice maintained with a limited food supply. Experientia (Basel) 35:603–604

Melnyk RB (1983) Accelerated circannual cycles in ground squirrels, *Spermophilus richardsonii*, kept in constant conditions. Can J Zool 61:1765–1770

Menaker M (1974) Circannual rhythms in circadian perspective. In: Pengelley ET (ed) Circannual clocks. Academic Press, New York, pp 507–518

Menaker M, Binkley S (1981) Neural and endocrine control of circadian rhythms in the vertebrates. In: Aschoff J (ed) Handbook of behavioral neurobiology, vol 4. Plenum, New York, pp 243–254

Merkel FW (1956) Untersuchungen über tages- und jahresperiodische Aktivitätsänderungen bei gekäfigten Zugvögeln. Z Tierpsychol 13:278–301

Merkel FW (1963) Long-term effects of constant photoperiods on European robins and whitethroats. Proc 13th Int Ornithol Congr Ithaca, pp 950–959

Mewaldt CR, Morton ML, Brown JL (1964) Orientation of migratory restlessness in *Zonotrichia*. Condor 66:377–417

Michael RP, Bonsall RW (1977) A 3-year study of an annual rhythm in plasma androgen levels in male rhesus monkeys (*Macaca mulatta*) in a constant laboratory environment. J Reprod Fertil 49:129–131

Michael RP, Zumpe D (1978) Annual cycles of aggression and plasma testosterone in captive male rhesus monkeys. Psychoneuroendocrinology 3:217–220

Michener GR (1984) Age, sex and species differences in the annual cycles of ground-dwelling *Sciurids*: implications for sociality. In: Murie JO, Michener GR (eds) The biology of ground-dwelling *Sciurids*. Univ Nebraska Press, Lincoln, pp 81–107

Miller AH (1960) Adaptation of breeding schedule to latitude. Proc 12th Int Ornithol Congr Helsinki 1958, pp 513–522

Moore-Ede M, Sulzman FM, Fuller CA (1982) The clocks that time us. Harvard Univ Press, Cambridge Mass

Moreau RE (1931) Equatorial reflections on periodism in birds. Ibis 1:553–570

Moreau RE, Wilk AL, Rowan W (1947) The moult and gonad cycles of three species of birds at five degrees south of the equator. Proc Zool Soc Lond 117:345–364

Morin LP, Fitzgerald KM, Rusak B, Zucker I (1977) Circadian organization and neural mediation of hamster reproductive rhythms. Psychoneuroendocrinology 2:73–98

Morrison P (1964) Adaptation of small mammals to the arctic. Fed Proc 23:1202–1206

Mrosovsky N (1970) Mechanism of hibernation cycles in ground squirrels: circannian rhythm or sequence of stages? Pennsylvania Acad Sci 44:172–175

Mrosovsky N (1974a) Hypothalamic hyperphagia without plateau in ground squirrel. Physiol Behav 12:259–264

Mrosovsky N (1974b) Comment. In: Pengelley ET (ed) Circannual clocks. Academic Press, New York, pp 161–163

Mrosovsky N (1975) The amplitude and period of circannual cycles of body weight in golden-mantled ground squirrels with medial hypothalamic lesions. Brain Res 99:97–116

Mrosovsky N (1977) Hibernation and body weight in dormice: a new type of endogenous cycle. Science 196:902–903

Mrosovsky N (1978) Circannual cycles in hibernators. In: Wang L, Hudson JW (eds) Strategies in cold: natural torpidity and thermogenesis. Academic Press, New York, pp 21–65

Mrosovsky N (1980a) Circannual cycles in golden-mantled ground squirrels: phase shift produced by low temperatures. J Comp Physiol 136:349–353

Mrosovsky N (1980b) Circannual cycles in golden-mantled ground squirrels: experiments with food deprivation and effects of temperature on periodicity. J Comp Physiol 136:355–360

Mrosovsky N (1985) Cyclical obesity in hibernators: the search for the adjustable regulator. In: Hirsch J, von Itallie T (eds) Recent advances in obesity research IV. Libey, London, pp 45–56

Mrosovsky N, Lang K (1971) Disturbances in the annual weight and hibernation cycles of thirteen-lined ground squirrels kept in constant conditions and the effects of temperature changes. J Interdiscip Cycle Res 2:79–90

Mrosovsky N, Lang K (1980) Body weights of garden dormice, *Eliomys quercinus*, kept in constant conditions for 2 years. Comp Biochem Physiol 67A:667–669

Mrosovsky N, Boshes M, Hallonquist JD, Lang K (1976) Circannual cycle of circadian cycles in a golden-mantled ground squirrel. Naturwissenschaften 6:298

Mrosovsky N, Melnyk RB, Lang K, Hallonquist JD, Boshes M, Joy JE (1980) Infradian cycles in dormice (*Glis glis*). J Comp Physiol 137:315–339

Muchlinski AE (1980) The effects of daylength and temperature on the hibernating rhythm of the meadow jumping mouse (*Zapus hudsonius*). Physiol Zool 53:410–418

Müller-Haeckel A (1975) Endogene Jahresperiodik der Blattbewegungen zweier *Oxalis*-Arten. Physiol Plant 35:236–242

Murton RK, Westwood NJ (1977) Avian breeding cycles. Clarendon, Oxford

Naumann J (1822) Naturgeschichte der Vögel Deutschlands. Fleischer, Leipzig

Nicholls TJ, Goldsmith AR, Dawson A, Chakraborty S, Follett BK (1985) Involvement of the thyroid gland in photorefractoriness in starlings. In: Follett BK, Ishii S, Chandola A (eds) The endocrine system and the environment. Japan Sci Soc Press, Tokyo. Springer, Berlin Heidelberg New York, pp 127–135

Nottebohm F (1981) A brain for all seasons: cyclical anatomical changes in song control nuclei in the canary brain. Science 214:1368–1370

Odum EP (1960) Lipid deposition in nocturnal migrant birds. Proc 12th Int Ornithol Congr Helsinki 1958, pp 563–576

Okada Y (1930) Study of *Euryale ferox* Salisb. V: On some features in the physiology of the seed with special respect to the problem of the delayed germination. Science Rep Res Inst Tohoku Univ 5:41–116

Palmgren P (1937) Auslösung der Frühlingszugunruhe durch Wärme bei gekäfigten Rotkehlchen, *Erithacus rubecula* L. Ornis Fenn 14:71–73

Payne RB (1972) Mechanisms and control of molt. In: Farner DS, King JR (eds) Avian biology. Academic Press, New York, pp 103–155

Pelletier J (1973) Evidence for photoperiodic control of prolactin release in rams. J Reprod Fertil 35:143–147

Pengelley ET (1968) Interrelationships of circannian rhythms in the ground squirrel, *Citellus lateralis*. Comp Biochem Physiol 24:915–919

Pengelley ET (ed) (1974) Circannual clocks. Academic Press, New York

Pengelley ET, Asmundson SJ (1969) Free-running periods of endogenous circannian rhythms in the golden-mantled ground squirrel, *Citellus lateralis*. Comp Biochem Physiol 30:177–183

Pengelley ET, Asmundson SJ (1970) The effect of light on the free-running circannual rhythm of the golden-mantled ground squirrel, *Citellus lateralis*. Comp Biochem Physiol 32:155–160

Pengelley ET, Asmundson SJ (1974) Circannual rhythmicity in hibernating mammals. In: Pengelley ET (ed) Circannual clocks. Academic Press, New York, pp 95–160

Pengelley ET, Asmundson SJ (1975) Female gestation and lactation as zeitgebers for circannual rhythmicity in the hibernating ground squirrel, *Citellus lateralis*. Comp Biochem Physiol 50A:621–625

Pengelley ET, Fisher KC (1957) Onset and cessation of hibernation under constant temperature and light in the golden-mantled ground squirrel, *Citellus lateralis*. Nature 180:1371–1372

Pengelley ET, Fisher KC (1963) The effect of temperature and photoperiod on the yearly hibernating behavior of captive golden-mantled ground squirrels (*Citellus lateralis tescorum*). Can J Zool 41:1103–1120

Pengelley ET, Kelly KH (1966) A "circannian" rhythm in hibernating species of the genus *Citellus* with observations on their physiological evolution. Comp Biochem Physiol 19:603–617

Pengelley ET, Asmundson SJ, Barnes B, Aloia RC (1976a) Relationship of light intensity and photoperiod to circannual rhythmicity in the hibernating ground squirrel, *Citellus lateralis*. Comp Biochem Physiol 53A:273–277

Pengelley ET, Asmundson SJ, Aloia RC, Barnes B (1976b) Circannual rhythmicity in a non-hibernating ground squirrel, *Citellus leucurus*. Comp Biochem Physiol 54A:233–237

Pengelley ET, Aloia RC, Barnes B (1978) Circannual rhythmicity in the hibernating ground squirrel, *Citellus lateralis,* under constant and hyperthermic ambient temperature. Comp Biochem Physiol 61A:599–603

Pengelley ET, Aloia RC, Barnes B, Whitson D (1979) Differential temporal behavior between males and females in the hibernating ground squirrel, *Citellus lateralis*. Comp Biochem Physiol 64A:593–596

Perdeck AC (1958) Two types of orientation in migrating starlings, *Sturnus vulgaris* L., and chaffinches, *Fringilla coelebs* L., as revealed by displacement experiments. Ardea 46:1–37

Perdeck AC (1964) An experiment on the ending of autumn migration in starlings. Ardea 52:133–139

Pernau FA von (1702) Unterricht was mit dem lieblichen Geschöpff, denen Vögeln, auch ausser dem Fang, nur durch Ergründung deren Eigenschafften und Zahmmachung oder anderer Abrichtung man sich vor Lust und Zeitvertreib machen könne. Nürnberg

Petter-Rousseaux A (1972) Application d'un système sémestriel de variation de la photopériode chez *Microcebus murinus* (Miller 1777). Ann Biol Anim Biochem Biophys 12:367–375

Petter-Rousseaux A (1975) Activite sexuelle de *Microcebus murinus* (Miller 1777) soumis a des régimes photopériodiques experimentaux. Ann Biol Anim Biochem Biophys 15:503–508

Phillips JA, Harlow HJ (1982) Long-term effects of pinealectomy on the annual cycle of golden-mantled ground squirrels, *Spermophilus lateralis*. J Comp Physiol 146:501–505

Pirson A, Göllner E (1953) Beobachtungen zur Entwicklungsphysiologie der *Lemna minor* L. Flora (Jena) 140:485–498

Pittendrigh CG (1960) Circadian rhythms and the circadian organization of living systems. Cold Spring Harbor Symp Quant Biol 25:159–184

Pittendrigh CS (1966) The circadian oscillation in *Drosophila pseudoobscura* pupae: a model for the photoperiodic clock. Z Pflanzenphysiol 54:275–307

Pittendrigh CS (1972) Circadian surfaces and the diversity of possible roles of circadian organization in photoperiodic induction. Proc Natl Acad Sci USA 69:2734–2737

Pittendrigh CS (1981a) Circadian systems: entrainment. In: Aschoff J (ed) Handbook of behavioral neurobiology, vol 4. Plenum, New York, pp 95–124

Pittendrigh CS (1981b) Circadian organization and the photoperiodic phenomena. In: Follett BK, Follett DE (eds) Biological clocks in seasonal reproductive cycles. Scientechnica, Bristol, pp 1–35

Pittendrigh CS, Daan S (1976) A functional analysis of circadian pacemakers in nocturnal rodents. V. Pacemaker structure: a clock for all seasons. J Comp Physiol 106:333–355

Pohl H (1968) Einfluß der Temperatur auf die freilaufende circadiane Aktivitätsperiodik bei Warmblütern. Z Vergl Physiol 58:364–380

Pohl H (1971) Circannuale Periodik beim Bergfinken. Naturwissenschaften 58:572–573

Pohl H (1972) Seasonal change in light sensitivity in *Carduelis flammea*. Naturwissenschaften 59:518

Priedkalns J, Bennett RK (1978) Environmental factors regulating gonadal growth in the Zebra Finch, *Taeniopygia guttata castanotis*. Gen Comp Endocrinol 34:80

Priedkalns J, Oksche A, Vleck C, Bennett RK (1984) The response of the hypothalamo-gonadal system to environmental factors in the zebra finch, *Poephila guttata castanotis*. Cell Tissue Res 238:23–35

Prosser RA, Satinoff E (1984) Suprachiasmatic nuclear lesions alter but do not eliminate circadian body temperature rhythms in rats. Fed Proc 43:907

Radford MH (1961) Photoperiodism and sexual activity in merino ewes II. The effect of equator light on sexual activity. Aust J Agric Res 12:147–153

Ralph CL, Harlow HJ, Phillips JA (1982) Delayed effect of pinealectomy on hibernation of the golden-mantled ground squirrel. Int J Biometeorol 26:311–328

Reiter RJ (1974) Circannual reproductive rhythms in mammals related to photoperiod and pineal function: a review. Chronobiologia 1:365–395

Reiter RJ (1980) The pineal and its hormones in the control of reproduction in mammals. Endocr Rev 1:109–131

Remmert HV (1965) Biologische Periodik. In: Gessner F (ed) Handbuch der Biologie. Akademische Verlagsgesellschaft Athenaion, Frankfurt am Main, pp 335–411

Reppert SM, Perlow MJ, Ungerleider LR, Mishkin M, Tamarkin L, Orloff DG, Hoffmann HJ, Klein DC (1981) Effects of damage to the suprachiasmatic area of the anterior hypothalamus on the daily melatonin and cortisol rhythms in the rhesus monkey. J Neurosci 1:1414–1425

Resende E (1947) Observacoes sobre ritmo endonomico vegetal em Portugal e suas colonias. I.O. ritmo endonomico annual e a floracao em Chorisia crispiflora H,B,U.K. Bull Soc Prtug Sci Nat 15:123–127

Richter CP (1978) Evidence for existence of a yearly clock in surgically and self-blinded chipmunks. Proc Natl Acad Sci USA 75(7):3517–3521

Robinson JE, Follett BK (1982) Photoperiodism in Japanese quail: the termination of seasonal breeding by photorefractoriness. Proc R Soc Lond B Biol Sci 215:95–116

Rougeot J (1961) Actions comparées des variations périodiques, annuelles et semestrelles de la durée quotidienne de l'éclairement sur les cycles des follicules des jarres courts de la toison des brebis limousines. Relations avec leur cycle de réproduction. Ann Biol Anim Biochem Biophys 1:385–402

Rougeot J (1962) Action du photoperiodisme saisonnier sur l'activité des follicules pileux de la toison des ovins. In: Thiers H, Cotte J (eds) Actualités de dermopharmacologie 4, Centre Europ de Dermapharmacologie, Lyon, pp 155–170

Rowan W (1926) On photoperiodism, reproductive periodicity and the annual migrations of birds and certain fishes. Proc Boston Soc Nat Hist 38:147–189

Rüppell W, Schütz E (1948) Ergebnis der Verfrachtung von Nebelkrähen (*Corvus corone cornix*) während des Wegzuges. Vogelwarte 15:30–36

Ruge U, Liedtke D (1951) Zur periodischen Keimbereitschaft einiger Malven-Arten. Ber Dtsch Bot Ges 64:141–150

Rusak B, Zucker I (1979) Neural regulation of circadian rhythms. Physiol Rev 59:449–526

Rutledge JT (1974) Circannual rhythm of reproduction in male European starlings (*Sturnus vulgaris*). In: Pengelley ET (ed) Circannual clocks. Academic Press, New York, pp 297–345

Saboureau M (1981) Environmental factors and regulation of the annual testicular cycle in a hibernating mammal: the hedgehog. In: INRA Publ. (Les colloques de l'INRA, 6), pp 319–337

Sansum EL, King JR (1976) Long-term effects of constant photoperiods on testicular cycles of white-crowned sparrows (*Zonotrichia leucophrys gambelii*). Physiol Zool 49:407–416

Sauer EGF (1957) Die Sternorientierung nächtlich ziehender Grasmücken (*Sylvia atricapilla, borin* und *curruca*). Z Tierpsychol 14:29–70

Saunders DS (1976) Insect clocks. Pergamon, New York

Saunders DS (1981) Insect photoperiodism. In: Aschoff J (ed) Handbook of behavioral neurobiology, vol 4. Plenum, New York, pp 411–447

Schleußner G (1984) Zur Rolle der circadianen Rhythmik bei der Beendigung der Photorefraktärphase des Stars (*Sturnus vulgaris* L.). Diplomarbeit Univ München

Schmidt-Koenig K (1975) Migration and homing in animals. Springer, Berlin Heidelberg New York

Schwab RG (1971) Circannian testicular periodicity in the European starling in the absence of photoperiodic change. In: Menaker M (ed) Biochronometry. Natl Acad Sci Wash DC, pp 428–447

Scott GW, Fisher KC (1970) The lengths of the hibernation cycles in mammalian hibernators living under controlled conditions. Pennsylvania Acad Sci 44:180–183

Scott GW, Fisher KC (1972) Hibernation of eastern chipmunks (*Tamias striatus*) maintained under controlled conditions. Can J Zool 50:95–105

Scott GW, Fisher KC (1976) Periodicity of hibernation of dormice (*Glis glis*) maintained under controlled conditions. Can J Zool 54:437–441

Segal E (1960) Discussion to the paper of AJ Marshall. Cold Spring Harbor Symp Quant Biol, pp 504–505

Serventy DL (1971) Biology of desert birds. In: Farner DS, King JR (eds) Avian biology, vol I. Academic Press, New York, pp 287–341

Serventy DL, Marshall AJ (1957) Breeding periodicity in western Australian birds; with an account of unseasonal nestings in 1953 and 1955. Emu 57:99–126

Simon SV (1914) Studien über die Periodizität der Lebensprozesse der in dauernd feuchten Tropengebieten heimischen Bäume. Jahrb Wiss Bot 54:71–187

Singh TP (1968) Effects of varied photoperiods on rhythmic activity of thyroid gland in a teleost, *Mystus vittatus* (Blach). Experientia (Basel) 24:93–94

Snow DW, Snow BK (1964) Breeding seasons and annual cycles of Trinidad land-birds. Zoologica (NY) 49:1–39

Sperlich A (1919) Über den Einfluß des Quellungszeitpunkts von Treibmitteln und des Lichtes auf die Samenkeimung von *Alectorolophus hirsutus* All., Characterisierung der Samenruhe. S.B. Akad Wiss Wien, Math-Naturwiss Klasse I, 128:477–500

Spruyt E, Maes L, Verbelen JP, Moereels E, DeGreef JA (1983) Circannual course of photomorphogenetic reactivity in etiolated bean seedlings. Photochem Photobiol 37:471–473

Stebbins RC (1963) Activity changes in the striped plateau lizard with evidence of influence of the parietal eye. Copeia 4:681–691

Stiles FG, Wolf LL (1974) A possible circannual molt rhythm in a tropical hummingbird. Am Nat 108:341–354

Sundararaj BI, Vasal S (1973) Photoperiodic regulation of reproductive cycle in the catfish, *Heteropneustes fossilis* (Bloch). Proc 4th Int Congr Endocrinol Wash DC 1972. Int Congr Ser 273. Excerpta Medica, Amsterdam, pp 180–184

Sundararaj BI, Vasal S, Halberg F (1982) Circannual rhythmic ovarian recrudescence in the catfish *Heteropneustes fossilis* (Bloch). In: Takahashi R, Halberg F, Walker CA (eds) Toward chronopharmacology. Pergamon Press, Oxford, pp 319–337

Sweeney BM (1969) Rhythmic phenomena in plants. Academic Press, New York

Thomson AL (1950) Factors determining the breeding seasons of birds: an introductory review. Ibis 92:173–184

Thwaites CJ (1965) Photoperiodic control of breeding activity in the southdown ewe with particular reference to the effects of an equatorial light regime. J Agric Sci 65:57–64

Turek FW (1978) Diurnal rhythms and the seasonal reproductive cycle in birds. In: Assenmacher I, Farner DS (eds) Environmental endocrinology. Springer, Berlin Heidelberg New York, pp 144–152

Turek FW (1983) Neurobiology of circadian rhythms in mammals. Bioscience 33:439–444

Turek FW, Campbell CS (1979) Photoperiodic regulation of neuroendocrine-gonadal activity. Biol Reprod 20:32–50

Turek FW, Gwinner E (1982) Role of hormones in the circadian organization of vertebrates. In: Aschoff J, Daan S, Groos G (eds) Vertebrate circadian systems. Springer, Berlin Heidelberg New York, pp 173–182

Tyshchenko VP (1966) Two-oscillatory model of the physiological mechanism of insect photoperiodic reaction (Russian). Zh Obshchei Biol 27:209–222

Vince-Prue D (1975) Photoperiodism in plants. McGraw Hill, London

Volkens G (1912) Laubfall und Lauberneuerung in den Tropen. Bornträger, Berlin

Wagner HO (1957) Vogelzug, Umweltreize und Hormone. Verh Dtsch Zool Ges 1957, pp 289–298

Wagner HO, Schildmacher H (1937) Über die Abhängigkeit des Einsetzens der nächtlichen Zugunruhe von der geographischen Breite. Vogelzug 8:18–19

Walker JM, Haskell EH, Berger RJ, Heller HC (1980) Hibernation and circannual rhythms of sheep. Physiol Zool 53:8–11

Wallraff HG (1960) Does celestial navigation exist in animals? Cold Spring Harbor Symp Quant Biol 25:451–461

Walter H (1968) Zur Abhängigkeit des Eleonorenfalken (*Falco eleonorae*) vom mediterranen Vogelzug. J Ornithol 109:323–365

Ward JR, Armitage KB (1981) Circannual rhythms of food consumption, body mass, and metabolism in yellow-bellied marmots. Comp Biochem Physiol 69A:621–626

Weise CM (1963) Annual physiological cycles in captive birds of differing migratory habits. Proc 13th Int Ornithol Congr Ithaca, pp 983–993

Wever RA (1979) The circadian system of man. Springer, Berlin Heidelberg New York

White HL (1936) The interaction of factors in the growth of *Lemna* IX. Ann Bot (Lond) 50:827–848

Wickings EJ, Nieschlag E (1980) Seasonality in endocrine and exocrine testicular function of the adult rhesus monkey (*Macaca mulatta*) maintained in a controlled laboratory environment. Int J Androl 3:87–104

Wieselthier AS, van Tienhoven A (1972) The effect of thyroidectomy on testicular size and on the photorefractory period in the starling, *Sturnus vulgaris*. J Exp Zool 179:331–338

Wingfield JC, Farner DS (1980) Control of seasonal reproduction in temperate-zone birds. Progr Reprod Biol 5:62–101

Witschi E, Miller RA (1938) Ambisexuality in the female starling. J Exp Zool 79:475–486

Wodzicka-Tomaszewska M, Hutchinson JSD, Bennett JW (1967) Control of the annual rhythm of breeding ewes: effect of an equatorial daylength with reversed thermal seasons. J Agric Sci 69:61 ff.

Wolff WJ (1970) Goal orientation versus one direction orientation in the teal, *Anas c. crecca* during autumn migration. Ardea 58:132–141

Wong CC, Döhler KD, Atkinson MJ, Geerlings H, Hesch RD, zur Mühlen A von (1983) Circannual variations in serum concentrations of pituitary, thyroid, parathyroid, gonadal and adrenal hormones in male laboratory rats. J Endocrinol 97:179–185

Zatzman ML, South FE (1981) Circannual renal function and plasma electrolytes of the marmot. Am J Physiol 241:R87–R91

Zimmermann JL (1966) Effects of extended tropical photoperiod and temperature on the dickcissel. Condor 68:377–387

Zink G (1977) Richtungsänderungen auf dem Zug bei europäischen Singvögeln. Vogelwarte 29 (Sonderheft):144–154

Zink G (1973–1981) Der Zug europäischer Singvögel I, II, III. Vogelzug, Möggingen
Zucker I (1979) Hormones and hamster circadian organization. In: Suda M, Hayaishi O, Nakagawa H (eds) Biological rhythms and their central mechanism. Elsevier North Holland, Amsterdam, pp 369–381
Zucker I (1985) Pineal gland influences period of circannual rhythms of ground squirrels. Am J Physiol 249:R111–R115
Zucker I (1986) Neuroendocrine substrates of circannual rhythms. Psychobiology of depression. Taos NM (in press)
Zucker I, Boshes M (1982) Circannual body weight rhythms of ground squirrels: role of gonadal hormones. Am J Physiol 243:R546–R551
Zucker I, Licht P (1983a) Circannual and seasonal variations in plasma luteinizing hormone levels of ovariectomized ground squirrels (*Spermophilus lateralis*). Biol Reprod 28:178–185
Zucker I, Licht P (1983b) Seasonal variations in plasma luteinizing hormone levels of gonadectomized male ground squirrels. Biol Reprod 29:278–285
Zucker I, Boshes M, Dark J (1983) Suprachiasmatic nuclei influence circannual and circadian rhythms of ground squirrels. Am J Physiol 244:R472–R480

Systematic Index

Ammosphermophilus leucurus 15, 19, 45, 73, 84
Ankistrodesmus braunii 17, 31
Anthrenus verbasci 13, 29, 44, 56–58
Antrozous pallidus 14, 22
Arachnothera longirostris 7

Bat, Namie's frosted, see *Vespertilio superans*
Beetle, carpet, see *Anthrenus verbasci*
Bishop, red, see *Euplectes orix*
Blackcap, see *Sylvia atricapilla*
Bobolink, see *Dolichonyx oryzivorus*
Brambling, see *Fringilla montifringilla*

Calidris alpina 2
Campanularia flexuosa 13, 31
Canary, see *Serinus canaria*
Capra domestica 16, 21, 23
Catfish, airsac, see *Heteropneustes fossilis*
Catostomus commersoni 13, 28
Cervus nippon 16, 20–22, 39, 41, 44, 46, 50, 51, 53, 54, 56, 63–65
Chaffinch, see *Fringilla coelebs*
Chiffchaff, see *Phylloscopus collybita*
Chipmunk
 alpine, see *Eutamias alpinus*
 eastern, see *Tamias striatus*
 least, see *Eutamias minimus*
 lodgepole, see *Eutamias speciosus*
 yellow pine, see *Eutamias amoenus*
Corvus corone 122
Crayfish, cave, see *Orconectes pellucidus*
Cricetus cricetus 15, 20, 54, 60, 64, 96
Crossbill, see *Loxia curvirostra*
Crow, carrion, see *Corvus corone*

Dactylis glomerata 17, 35
Deear, Sika, see *Cervus nippon*
Dickcissel, see *Spiza americana*
Digitalis lutea 17, 32, 33
Dioch, red-billed, see *Quelea quelea*
Dolichonyx oryzivorus 18
Dormouse, common, see *Glis glis*
 garden, see *Eliomys quercinus*
Drosophila pseudoobscura 29

Eliomys quercinus 20, 36, 40
Erithacus rubecula 36, 37
Euplectes orix 18
Eutamias alpinus 16, 100
Eutamias amoenus 16, 100

Eutamias minimus 16, 19, 100
Eutamias speciosus 16, 100

Falco eleonorae 2
Falcon, Eleonora's, see *Falco eleonorae*
Ferret, see *Mustela putorius*
Ficedula albicollis 14, 40, 102, 109, 129
Ficedula hypoleuca 40, 102, 109, 129
Flyatcher, collared, see *Ficedula albicollis*
 pied, see *Ficedula hypoleuca*
Fragaria vesca 33, 34
Fringilla coelebs 14, 42
Fringilla montifringilla 18
Fundulus grandis 82

Gasterosteus aculeatus 36
Glis glis 16, 19, 21, 35, 36, 40, 43, 56, 60, 75, 76
Goat, see *Capra domestica*
Gratiola officinalis 17, 33
Ground squirrel, antelope, see
  Ammospermophilus leucurus
 belding, see *Spermophilus beldingi*
 California, see *Spermophilus beecheyi*
 golden-mantled, see *Spermophilus lateralis*
 Mohave, see *Spermophilus mohavensis*
 Richardson, see *Spermophilus richardsoni*
 rock, see *Spermophilus variegatus*
 round-tailed, see *Spermophilus tereticaudus*
 thirteen-lined, see *Spermophilus tridecemlineatus*

Hamster, European, see *Cricetus cricetus*
 golden, see *Mesocricetus auratus*
 Turkish, see *Mesocricetus brandti*
Helix aspera 17, 31, 56
Hermit, long-tailed, see *Phaethornis superciliosus*
Heteropneustes fossilis 17, 29, 45
Hypericum perforatum 17

Jumping mouse, meadow, see *Zapus hudsonius*
 western, see *Zapus princeps*

Kingfish, Gulf, see *Fundulus grandis*

Lanius collurio 108, 124
Lemna minor 13, 17, 31, 32
Lemur, mouse, see *Microcebus murinus*
Limax flavus 13, 30, 31, 44, 47
Lizard, striped plateau, see *Sceloporus virgatus*
Lonchura punctulata 14, 45
Loxia curvirostra 14

Macaca mulatta 14, 18, 19, 22
Marmot, yellow-bellied, see *Marmota flaviventris*
Marmota flaviventris 19, 20, 107
Marmota monax 16, 19, 21, 49
Mesocricetus auratus 78, 82, 96
Mesocricetus brandti 20, 40, 89
Microcebus murinus 18, 54, 64
Microtus agrestis 20
Monkey, rhesus, see *Macaca mulatta*
Mouse, house, see *Mus musculus*
Munia, spotted, see *Lonchura punctulata*
Mus musculus 42
Mustela putorius 20, 94–96
Mystus vittatus 17, 29

Orconectes pellucidus 13, 30, 31
Ovis aries 16, 20, 22, 23, 44, 46, 55–57, 60, 61, 64, 84, 94–96
Oxalis 35

Parus cristatus 13
Pekin duck 35–37
Perognathus longimembris 16, 21
Phaethornis superciliosus 87, 88
Phaseolus vulgaris 17
Phylloscopus collybita 39, 101–105, 108, 110, 111, 114, 116, 121
Phylloscopus trochilus 13, 24, 37, 39, 101–105, 108–112, 114, 116, 120, 121, 127, 128
Pocket mouse, little, see *Perognathus longimembris*
Potentilla molissima 17
Pseudemys scripta 18, 28

Quelea quelea 18, 39

Robin, European, see *Erithacus rubecula*

Salmon, Baltic, see *Salmo salar*
Salmo salar 17
Salmo trutta 17
Sandpiper, red-backed, see *Calidris alpina*
Saxicola torquata, 47, 102, 109
Sceloporus virgatus 13, 28
Serinus canaria 97
Sheep, see *Ovis aries*
Shrike, red-backed, see *Lanius collurio*
Sparrow, white-crowned, see *Zonotrichia leucophrys*
  white-throated, see *Zonotrichia albicollis*
Spermophilus beecheyi 15, 58, 107
Spermophilus beldingi 15, 100, 101
Spermophilus lateralis 5, 11, 12, 15, 19, 21, 39, 41–45, 47, 58–60, 75, 79, 80, 83, 84, 88–90, 94, 95, 100, 101, 107
Spermophilus mohavensis 19, 100, 107
Spermophilus richardsoni 36

Spermophilus tereticaudus 100, 107
Spermophilus tridecemlineatus 15, 19, 84, 85, 107
Spermophilus variegatus 107
Spiderhunter, little, see *Arachnothera longirostris*
Spirodela polyrrhiza 17
Spiza americana 18, 39, 108
Stachyris erythroptera 7
Starling, European, see *Sturnus vulgaris*
Sterna fuscata 7
Stickleback, three-spined, see *Gasterosteus aculeatus*
Stonechat, see *Saxicola torquata*
Sturnus vulgaris 14, 18, 27, 39, 41, 44, 45, 47, 50–52, 54, 56, 57, 61–66, 68, 70, 71, 74, 75, 77, 78, 80, 81, 84, 85, 89, 91–94, 97, 122, 123
Sucker, white, see *Catostomus commersoni*
Sylvia atricapilla 13, 24–26, 39, 41, 45, 84, 108, 113, 115, 116, 119
Sylvia borin 13, 18, 24–27, 39, 41, 44, 45, 54, 70, 71, 73, 83, 85–87, 117, 118, 125, 126
Sylvia cantillans 14, 115, 116
Sylvia communis 18, 38–40, 108, 112, 115
Sylvia hortensis 124
Sylvia melanocephala 14, 115, 116
Sylvia nisoria 115, 116, 124
Sylvia sarda 14, 115, 116
Sylvia undata 14, 115, 116

Taeniopygia guttata 3
Tamias striatus 16, 19, 21, 22, 39, 42, 43, 47, 73
Tern, sooty, see *Sterna fuscata*
Tit, crested, see *Parus cristatus*
Tree-babbler, red-winged, see *Stachyris erythroptera*
Trout, brown, see *Salmo trutta*
Turtle, red-eared, see *Pseudemys scripta*

Vespertilio superans 19, 89
Vole, field, see *Microtus agrestis*

Warbler, barred, see *Sylvia nisoria*
  Dartford, see *Sylvia undata*
  garden, see *Sylvia borin*
  Marmora, see *Sylvia sarda*
  orphean, see *Sylvia hortensis*
  Sardinian, see *Sylvia melanocephala*
  subalpine, see *Sylvia cantillans*
  willow, see *Phylloscopus trochilus*
Whitethroat, see *Sylvia communis*
Woodchuck, see *Marmota monax*

Zapus hudsonius 20, 40
Zapus princeps 16, 21
Zebrafinch, see *Taeniopygia guttata*
Zonotrichia albicollis 82
Zonotrichia leucophrys 6, 14, 18, 40, 45

Subject Index

Australian desert birds 3

Bünning's hypothesis 76
Breeding cycles
  10-months 7
  yearly 6, 99–101, 105–106

Circadian rhythms
  affected by cirannual rhythms 76–82
  comparison with cirannual rhythms 8, 47–48, 67–68
  external coincidence model 76–78
  generating cirannual rhythms 70–76
  internal coincidence model 78–82
  leaf-movements 35
Circannual rhythms
  adaptive significance 4–8, 99–129
  antler replacement in deer 16, 21–22, 39, 41, 46, 51, 53–54, 56, 61, 64
  arresting rhythmicity 91–92
  atypical cases 35–38
  birds 5–7, 13–14, 18, 24–28, 36–42, 44–45, 47, 50–52, 54, 61–64, 66, 70–75, 77–89, 91–94, 96, 101–106, 108–129
  blood amino acid nitrogen 18
  blood lipid phosphorus 18
  body weight 11–16, 18–22, 24–25, 36–40, 42–45, 49, 54, 58–60, 63–66, 73, 76, 79–80, 82–84, 86–90, 94–96, 100–104, 107, 110–113, 116, 118
  brain noradrenalin 20
  catalase activity 17
  comparison with circadian rhythms 47–48, 67–68
  definitions 8–9
  degree of persistence 39–41
  diapause in insects 13, 29, 44, 56–57
  dry matter production in plants 13, 31–32
  fish 13, 17, 28–29, 36, 45
  food consumption 11–12, 14–17, 19, 21–22, 31, 36, 73, 88
  food preference 18
  free running 8–9, 11–32, 39–48
  frequency demultiplication of circadian rhythms 70–76
  general properties 8–48
  germination of seeds 17, 31–35

gonadectomy 89, 94
growth in animals 13, 31
growth in plants 17, 31
heat resistance in seeds 17
hibernation 5, 11–12, 15–16, 19–21, 35–37, 39–45, 47, 58–59, 83, 88–90, 95, 100–101, 106–108
history 5–6
hypothalamic lesions 89–90, 95–96
innateness 47
interactions with circadian rhythms 69–82
internal dissociation 85–88
invertebrates 13, 29–31, 44, 47, 56
locomotor activity 13, 16, 18–20, 28, 73, 79–81
mammals 5, 11–12, 14–16, 18–24, 35–36, 39–49, 51, 53–61, 64, 67, 73, 75–76, 78–80, 83–84, 88–91, 94–97, 100–101, 105–108
mechanisms 69–97
milk production in goats 16, 21, 23
molt 7, 13–15, 18–19, 24–26, 27, 30–31, 36, 39–41, 44–46, 50–52, 54–55, 58, 61–66, 70–71, 84–89, 91–93, 101–103, 105, 109–113, 127–128
multi-oscillator system 82–90
nitrate reduction in plants 17, 31
ontogeny 47
orientation in birds 125–126
oscillator model 8, 131–133
osmotic value in plants 17
oxygen consumption 19
pattern of migration in birds 114–125
perennial organ formation in plants 17, 35
permissive conditions 9, 39–40, 81–93
photomorphogenetic activity in plants 17, 35
photoperiod effects on freerunning period 44–46
pinealectomy 74–75, 94–95
plants 7–8, 13, 17, 31–35
plasmolysis in plants 17
range of entrainment 61–66
range of period values 41–42
renal function 20
reproductive parameters 6–7, 13–23, 26–40, 44–47, 50–52, 54–58, 60–66, 70–97, 100–101
reptiles 13, 28
skin coloration in fish 17, 28
synchronisation 49–68

153

Circannual rhythms
- temperature effects on freerunning period 42–44
- thermoregulation 13, 19, 28
- thyroid activity 17, 29, 36
- timing of migration in birds 108–113
- transients 41, 49
- tropical animals 6–7, 105–106
- urinary volume 20
- water intake 15–16, 19, 22, 36, 73, 84
- zeitgebers 49–61
- zugunruhe 13–14, 18, 24–25, 39, 45, 47, 54, 84, 86–87, 101–103, 105, 108–128

Endogenous rhythms
- criteria for 8–9
- free running 8–9, 11–32, 39–48
- terminology 131–133

Frequency demultiplication 70–76, 131–133

Genetics of circannual programs 113, 118

Hormones
- circadian patterns 82
- cirannual patterns 14, 16–20, 36, 65, 82
- effects on circadian rhythms 79–80
- possible components of circannual rhythms 94–95

Infradian rhythms 35–38, 40–41

Masking 65, 92

Photoperiod
- as factor modifying circannual programs 125–128
- as permissive condition 9, 39–40, 84, 91–93
- as proximate factor 4
- as zeitgeber 50–58, 61–68
- effects on circannual period 44–46

Pineal
- effects on circadian rhythms 74
- effects on circannual rhythms 74, 94–95

Proximate factors
- controlling annual rhythms 2–5
- hierarchical organization 4–5

Self sustaining rhythms 61–66, 131–133
Sequence of stages 82–90
Social factors
- as zeitgeber 60–61
- effects on circannual period 47
Suprachiasmatic nucleus
- effects on circadian rhythms 75, 95
- effects on circannual rhythms 75
Synchronization of circannual rhythms
- at different latitudes 66
- by frequency demultiplication 131–133
- phase-relationship to zeitgebers 63–66
- phase-response curves 50, 56, 59–60
- range of entrainment 61–66
- transients 49, 62
- zeitgebers 49–61

Temperature
- effects on circannual period 42–44
- permissive condition 40, 84
- proximate factor 3
- zeitgeber 58–60, 67–68
Timing, significance of
- circannual clock 4–7, 99–129
- equatorial breeders 6–7
- hibernation 100–101, 106–108
- long-distance migrants 5–6, 101–106, 108–129
- migratory direction 125–126
- migratory distance 114–125

Ultimate factors in the control of annual rhythms 1–2

Zeitgebers
- criteria for a stimulus being a 50
- of circannual rhythms 49–61

DATE DUE

DEC 1 1999 RETURNED SEP 1 4 1999			

DEMCO 38-297